FOR YEARS, READERS OF BC OUTDOORS MAGAZINE HAVE CLIPPED and collected the fly of the month and fly tying columns from each issue and repeatedly asked, "When are you going to do a book?"

As BC *Outdoors* approaches 60 years of publishing an outdoors magazine for and about British Columbia's great adventure seekers, we finally have an answer — the time is now.

The flies selected best represent the evolution of fly tying in this province and the innovative ability of B.C. fly fishers to adapt a fly to fish in the varied and specific waters that make this province a world renowned fishing adventure destination.

We hope you enjoy this book as much as we enjoyed putting it together for you.

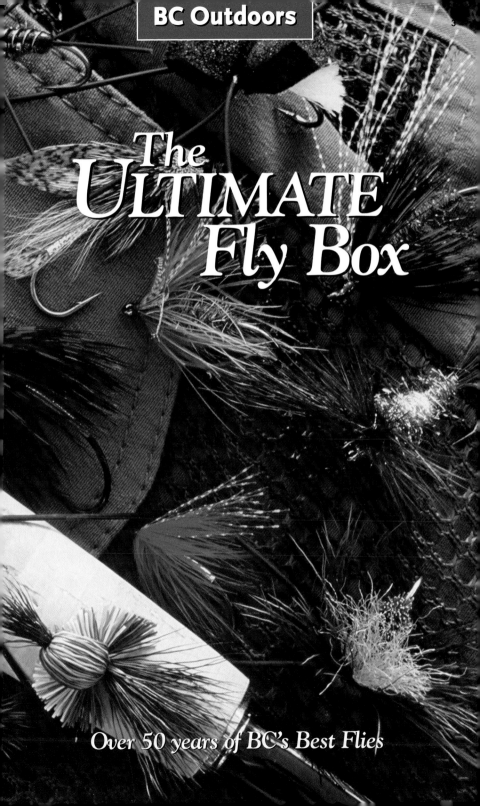

ACKNOWLEDGEMENTS

This book came about through the collaborative efforts of many dedicated members of the BC Outdoors community:
Scott Baker-McGarva, Berry's Bait and Tackle, Ehor Boyanowsky, Brian Chan, Hermann Fischer, Greg Flinn, Ian Forbes, Janet Genders, Tom Johannesen, David Kimble, David Lambroughton, Harold Lohr, Jim McLennan, the Mid-Island Castaways Fly Tying Club (Shawn Brown, Harvey DeBuc, Glen Gordon, Courtney Ogilvie, Bill Pollard, Bob Weir), Outback Fly Shop, Dave Owen, Lisa Penz, D.C. Reid, Ian Ricketson, Ruddick's Fly Shop, Phil Rowley, Ed Smith, Matthew Symons, Mark Yelic.

And to you, the reader, whose repeated requests for a book were the driving force behind all our efforts. Thank you.

CREDITS

Design by Sonya Fuller and Kelly Craft
Edited by Tracey Ellis
Editorial Assistants: Holly Munn and David Webb

Copyright ©2004 OP Publishing Ltd.
All rights reserved. No part of this publication may be reproduced or used in any form or by any means without the prior permission of OP Publishing Ltd. 1080 Howe Street, Suite 900, Vancouver, BC V6Z 2T1. Phone 604-606-4644. www.oppublishing.com

Printed and bound in Canada

National Library of Canada Cataloguing in Publication
The ultimate fly box : over 50 years of BC's best flies.
Edited by Tracey Ellis.
"Presents fly patterns that were featured in BC Outdoors magazines over the past 59 years".
Includes index.
ISBN 1-896373-66-6

1.†Flies, Artificial—British Columbia—History—20th century.
I.†Ellis, Tracey, 1967- II. BC outdoors.

SH451.U47 2004   688.7'9124'0971109045   C2004-900044-6

Table
of Contents

The innovative nature of British Columbia fly fishers has seen the adaptation of many traditional fly patterns to take advantage of the varied fishing opportunities unique to this province.

Foreword 7

Tips on Fly Fishing
    from an 'Oldtimer' 9

Nymphs 11

Emergers 67

Knots 75

Dry Flies 83

Salmonids 121

Contributors 186

Index 189

Foreword

British Columbia has a rich history when in comes to fly fishing and the development of fly patterns. Angling exploits that were written as far back as the late 1800s expounded on the virtues of the fabulous fishing opportunities that existed throughout the then still relatively inaccessible province. By the early 1900s word about the fabulous fishing in B.C. was getting out to serious anglers from around the world. Notable anglers such as Bryan Williams, Tommy Brayshaw and Bill Nation created many fly patterns that were well ahead of the traditional thinking around fly design for that time period. These innovative patterns laid the foundation for successive generations of anglers and tiers throughout the fly tying world.

This book, The Ultimate Fly Box presents fly patterns that were featured in BC Outdoors Magazines over the past 59 years. This collection includes more than 150 patterns right from the early era of B.C. fly fishing to the most current designs for salt water, rivers and lakes. Most importantly, these patterns catch fish. Many anglers will have heard of some of these patterns but may have not known any history about the fly or how the fly was tied. The evolution of an effective fly pattern typically sees many iterations or refinements to an original idea before the final design is reached. These are time tested flies that worked 50 years ago or this past fishing season and are worthy to be included in your fly box.

There are flies for cutthroat and salmon in both salt and fresh water as well as dry flies, Emergers, Pupa and Nymphs for trout and char in both lakes and rivers. And finally, proven effective patterns for both winter and summer run Steelhead are provided for those who seek out these special fish. A detailed materials list that accompanies each fly pattern is further enhanced with tying tips or suggestions on how to fish the fly. Readers will be exposed to both traditional tying materials as well as the latest synthetics, which complement many of the innovative tying techniques employed by the tiers of these patterns. The diversity of patterns included in this book will stimulate the creative juices of both beginner and advanced fly tiers.

The Ultimate Fly Box offers more than enough fly patterns to prepare you to fish not only those local waters you have always wanted to visit but many other destinations throughout North America and abroad. Enjoy reading about these patterns and more importantly, tie some up and enjoy the fruits of your labour.

— Brian Chan

Ah, the good ol' days...

Tips on Fly Fishing from an 'Oldtimer'

By F.W.L.

SUMMER 1946
VOL. 2 NO. 2
ISSUE PRICE: 35¢

The oldtimer looked through my flybook and smiled up at me. "Plumb purty, ain't they?" he chuckled. "I allus claimed that these purty flies caught one fish once, and that's the sucker that bought 'em."

I looked at my flies with a new interest. They had cost me a-plenty. I had dozens of them — all shapes and sizes, and colours. There was every colour, from the brilliant yellow of a "Mexican Killer" to the languorous purple of a "Northern Dream."

"What flies do you use?" I asked the oldtimer.

He took off his battered hat and showed me two flies stuck in the hatband.

"Them two'll catch me all the fish I want. This here's a 'Royal Coachman,' on a number 10 hook. T'other's a 'Zulu.' When she's kinda overcast I use the Coachman, when the day is bright and shinin' I use the 'Zulu.' I can pack 'em around in my hat all the flies I need."

"There must be other flies that can kill fish," I urged the oldtimer. "Some of my flies must be good."

He nodded his head. "Sure," he admitted, "there are lots of flies that can catch fish, but up here in Cariboo the fish are a trifle intelligent. The never did see a real fly with a three-inch wing spread, 'specially red wings that had a yellow body. Mebbe there are such flies in the hot countries or South America or Africy, or some place like that, but not here in Cariboo. Heck an' dangnation a fly oughtto look like a fly. That's what I think, and that's what the fish think, too."

Now this oldtimer was a renowned fisherman. I realized that he was pulling my leg, I further realized that he had a lot on the ball when it came to fishing lore, so instead of turning in for the night, I started to fish for information.

"What flies would you advise a tourist to bring with him if he was intending to catch fish up here?" I asked.

"Depends on the tourist," he grunted. "If he aims to catch fish, he should bring a few 'Royal Coachmen,' some 'Montreals,' Black Gnats, Zulus and mebbe some Bumble-Bees. If he has ever done any fishing at all, he'll know enough to use bright flies on a dark day and the dark flies on a bright day.

"Let us assume he knows the rudiments of the game," I suggested. "Let us presume he is a fisherman from the Southern States and used to catching bass or whatever they catch down there. Now he wants to try his hand at fly fishing, or maybe trolling, in the North. Suppose he asked you what tackle to bring, what would you tell him?"

"Well," the oldtimer scratches his head, "I already told you about the most important flies. If this here tourist was going to fish in fast water he'd want to have a couple of Silver Doctors handy, I guess. If he was coming up here early, say in May or the first part of June, he'd want an Evening Fly or else a Queen of the Waters, or, if he is loaded with money, he'd better bring a dozen of both."

"Would he need any special sized hook?" I queries.

"Ah!" the oldtimer snorted and shook his head. "Most of these dudes use a hook that would do for a meat hook. Half the fish in the country get sprained jaw muscles trying to get the darned things into their mouths. A number 6 is plenty large enough, but an 8 or 10 is best."

"What about dry flies?" I prodded gently, "what about them?"

"They are alright for them as likes 'em," chuckled the oldtimer. "But they don't get the fish a wet fly will — at least, they don't for me. In fact, I'd sooner use a spoon or a spinner any day than mess around with a flock of these here dry flies."

"Some folks use nothing else," I told him.

"Let 'em," he retorted. "Me, I never have any luck with 'em."

"Well, what about spoons?"

"Spoons are alright, too," said the oldtimer, stifling a yawn. "There is one spoon that should be taken off the market. It is a killer, a great chunk of jewelry that fish can't seem to resist. I mean the 'Davis Lake Gang Troll.' It is the clear rig for a man that want to catch a lot of small fish on a hand line. I used one up in Stuart Lake and caught so many fish I was ashamed of myself."

"Is there any type of spoon that can be used with a fly rod?" I asked.

"Yup," said the oldtimer, sleepily, "there sure the heck is. It's called the 'Gibbs Fresh and Salt Troll,' the finest spoon that was ever turned out. You can cast with it or troll with it. This spoon is so light that there is no drag on the line at all. A fellow can surely have sport with one of these spoons. You can catch large Rainbows with it. You see, Rainbows are cannibals and this F.S.T. looks for all the world like a baby Rainbow that has lost its mother."

The oldtimer pulled his sleeping bag out of the canoe and started to unroll it. "Tell this tourist that size 1 or 2 is the best lure for Rainbows, that is if he is going to use this Gibbs spoon I was telling you about. If he wants to catch himself a few char, he can get the same spoon in a larger size. G'night."

The oldtimer snored so loudly that, tired though I was, I could hardly sleep and through my head a Dark Montreal chased a Royal Coachman in a giddy whirl of colour.

Nymphs

Nymphs

Nymphs

Nymphs are the aquatic form of various insect species such as: Mayflies, Stoneflies, Dragonflies and Damselflies. They all have six legs, tails, wing pads and a noticeable head. They spend a year or more living near the bottom of rivers and lakes. In lakes they are more likely to be found closer to shore or on shoals. Most feed on vegetation, but some Dragonfly nymphs and Stonefly nymphs are predaceous. Stonefly nymphs need the well oxygenated water of streams to live. Although they can swim, they prefer to crawl along the bottom, and they move onto shore before hatching into adults. Dragonfly nymphs usually crawl, but they have the unique ability to expulse water through their abdomen and move like a little jet. They use this ability to capture other insects or escape capture themselves. Damselfly nymphs either crawl or swim with a unique side to side wiggling motion. Mayflies are the most diverse of the group. Depending on the species they live in a variety of locations, from slow moving streams or lakes to turbulent rivers. They cannot survive in polluted water. Some Mayfly nymphs prefer rich, mud bottom streams, and they swim similar in fashion to the Damselfly nymph. Some Mayfly nymphs are small and flat, so they can craw around the rocks of tumbling streams without being washed away. Other Mayfly nymphs, like the Baetis, are excellent little swimmers and can dart around like minnows. Most nymphs are mottled in colouration and blend in with their surroundings. The same species can vary greatly from lake to lake.

Fly tiers need to vary their colour combinations depending on where the fly is to be used. Tiers need to design their patterns to duplicate the various nymphs' size, shape and colour. An understanding of the various nymphs' movements goes a long way in choosing how to fish the imitation. Dragonfly nymph patterns can be trolled quite quickly, but a similar retrieve would be ineffective with a Damselfly pattern. Stonefly nymph patterns need to be kept right on the bottom. Mayfly nymph patterns can be fished with a variety of methods.

THE DEXHEIMER SEDGE

SEPTEMBER 1984

DESPITE THEIR RAMPANT POPULARITY WITH trout, most fly fishers don't view Scuds or at least using Scud patterns with the same appeal. This lack of enthusiasm is due in part to the limited success many fly fishers experience using Scud patterns. With the sheer volume of Scuds present in some lakes it seems a needle-in-a-haystack proposition. Yet throughout the open water season Scuds run neck and neck with Chironomids as the most popular food sources. During windy conditions Scud patterns are ideal choices. Look for rocky exposed shorelines and anchor a cast length from shore. Using a floating line and long leader in conjunction with a weighted pattern place a cast with in a foot of the bank. Begin a steady paced hand twist retrieve. Scatter pauses throughout the retrieve to simulate rest periods while allowing the pattern to sink down the slope of the shoreline. Expect the take within the first 10 feet as trout forage amongst the windswept rocks pouncing on the myriad of food sources disturbed by the agitated water. Trout hooked using this strategy always give a good account of themselves as the shallow water offers little escape other than long runs and aerobatic displays.

THE DEXHEIMER SEDGE

HOOK: #10 to #18 Mustad 3906
BODY: Yellowish-lime green wool
RIBBING: One fibre of peacock herl
HACKLE: Soft brown hackle
WING: Small, white tipped, red-brown mallard drake breast feather

INSTRUCTIONS: Lay a base of thread on the hook and tie in one peacock herl fibre, then a length of body wool at the bend. Wind on a compact body of wool to the hook eye and follow with an open spiral of the herl, forming a rib. Tie in a soft brown hackle feather and wind on three times, keeping it sparse. Then pull the hackle fibres below the hook and lock in place with a few wraps, forming a beard that veils the hook. Select a white-tipped brown mallard drake breast feather for the wing. Strip off the basal fibres until what remains will extend past the body but not beyond the bend of the hook. Tie it in so that the feather lies closely along the hook shank, cupping over the body.

THE CAVERHILL NYMPH

NOVEMBER/DECEMBER 1984

October is a magical time of the year to be fly fishing stillwaters. The crowds and noise of summer are long gone. The air is cool and brisk and many mornings begin with the breaking of ice. Deciduous trees are in full splendour providing for spectacular photographs. But perhaps best of all, trout are on the prowl knowing that winter is weeks, not months, away. Cruising the shallows looking for targets of opportunity trout respond to anything that resembles food. Scuds, Leeches, Bloodworm and immature Dragon Nymph patterns all work. The consummate opportunists, trout take advantage of all methods of harvest. In the late fall, aquatic vegetation is at its highest and is now beginning its annual decline. Diving birds, such as coots, forage amongst the vegetation diving down ripping clumps of vegetation free, dragging it to the surface for final consumption. This activity disturbs all manner of food and trout are only too willing to glide through picking off Scuds, Dragons and Chironomid larva as they tumble to regain their bearings. Casting both suggestive and imitative patterns amongst the grazing coots on intermediate or slow sinking lines yields impressive results.

THE CAVERHILL NYMPH

HOOK: #4 to #10 Mustad 9672 or 79580 (for trout), #2 to #6 Partridge Salmon (for steelhead)
BODY: Dubbed black seal fur
RIBBING: Fine oval silver tinsel
HACKLE: Muddy natural black, soft
HEAD: Peacock herl

INSTRUCTIONS: At a point over the barb of the hook, tie in a length of fine oval silver tinsel. Dub on a body of black seal fur using a dubbing loop painted with sticky wax over the rear half of the hook shank. Tie off the dubbing and wind the ribbing forward in three or four open spirals then tie it off as well. Tie in a full, soft black hackle feather and wind it several times around the waist of the fly before tying off. Tie in a peacock herl feather and wind the tying thread to the eye, following with the herl. Tie off the herl at the eye. Reinforce the herl head by counter-winding back over the herl head, making a wrap and returning to the eye. Whip finish.

THE DAMSELFLY NYMPH

JUNE 1985

The distinct sinusoidal swimming motion of a Damselfly Nymph provides a unique challenge for the fly fisher. Early in the hatch, many patterns duplicating the size and colour of the mature Nymphs should work. But as the hatch progresses trout develop a keen eye towards the Damsel's specific swimming motion and distinct slender profile. Material selection become key to consistent success. Stillwaters provide little in the way of current to animate a fly so patterns must possess inherent movement. Soft, subtle materials such as Marabou, aftershaft feathers and rabbit fur are ideal candidates to mimic the natural Nymphs. At rest or while the fly sinks these materials pulse and breath, coming to life on their own. Takes on the drop are a sure sign of trout approval. Successful patterns are skinny. Keep in mind that the widest point on the natural Nymph is its head and prominent eyes. Incorporating weight in the form of lead wire substitute or metal beads supplies additional motion. Loop knots such as the Non-Slip Loop or Duncan Loop in conjunction with a straight-eye hook provides supplemental behaviour assistance and motion.

THE DAMSELFLY NYMPH

HOOK: #10 Mustad 94840 or equivalent

THREAD: Green

TAIL: Green filoplume or olive Marabou

BODY/HEAD: Green filoplume or Marabou

HACKLE: Dyed olive Hungarian partridge breast feather

EYES: 60-pound-test monofilament, ends melted until ball-shaped

INSTRUCTIONS: Tie in the eyes using figure-eight wraps. Wind the tying thread to the bend and tie in a long, fluffy tail. Tie in a clump of Marabou or a filoplume and wrap it forward to the eyes, adding more materials as necessary. Tie in the hackle feather behind the eyes and continue winding the body materials forward, using figure-eight wraps around the eyes. Wrap the thread back to the hackle. Wind the partridge feather as you would a neck hackle, tie off and whip finish.

McLEOD'S CHRONOMID

MARL SHRIMP

HOOK: #10 to #20 Tiemco TMC 200

TAIL: White Antron yarn

BODY: Black, brown, olive or grey Hairtron dubbing, colour to match naturals

RIBBING: Silver wire

WING CASE: Eight to 10 strands of pheasant back fibres

THORAX: White or grey natural ostrich

GILLS: White or grey ostrich

INSTRUCTIONS: Tie in an extremely sparse and short tail of white yarn and a section of silver wire at the hook bend. Using a dubbing loop smeared with sticky wax wind on a body of Hairtron, tapered to thin at tail to fattest about two-thirds up the hook. Tie off body and wind on about four evenly spaced wraps of wire ribbing, tie off. Tie in wing case material by the butts. Tie in an ostrich feather and spin in the thread, winding forward to build the thorax then tie off. Tie an ostrich feather in front of the thorax, wrap two or three times and tie off. Pull wing case strands over the back of the fly and tie down just behind the eye, whip finish and glue.

HOOK: Tiemco 2457 #14 or #16 or equivalent

TAIL: Fine-textured dubbing in pale morning dun yellow to olive green

BODY: Same as tail

RIB: Fine gold wire

INSTRUCTIONS: Tie in a short tuft of dubbing approximately three millimetres long for tail. Tie in fine gold wire at the hook bend. Wax thread and spin on a slender length of dubbing. Wind forward to the eye of the hook and tie off. Wind gold wire in the opposite direction that the dubbing was wound to form ribbing. Tie off head. Pick out dubbing on the fly's underside to create the legs. Trim the dubbing on the back and sides to form the shrimp shape.

CASE CADDIS

NOVEMBER/DECEMBER 1985

Most sub surface presentations on rivers and streams dictate pattern presentation on or near the bottom. Most areas of any river or stream can be successfully covered using a floating line, nine- to 12-foot leader, strike indicator and split shot or other external weight as necessary. Deep water in conjunction with swift current presents a unique challenge that taxes the traditional floating line approach. The Brooks Method is a little used but effective approach developed by the late Charles Brooks to probe deep swift runs. Using a fast sinking type three line Charles Brooks plumbed deeper reaches using his Nymph patterns with great success. Starting at the head of the run make a short cast less than 20 feet quartering upstream. Immediately raise the rod tip to keep all excess fly line off the water allowing the pattern and line to sink. Follow the sinking line through the drift keeping the rod raised to the 11 o'clock position. Once the drift is complete allow the fly line to swing below. Let the line hang below for a few seconds and then strip in the slack to recast and induce any trout that may have followed the fly. Expect a take at any point in the drift — be alert when the fly line is in front and below prior to swinging, as this is when the fly is at its deepest. Be prepared for firm takes when using the Brooks Method.

CASE CADDIS

HOOK: #6 3X long

UNDERBODY: Silver tinsel chenille over lead wire

BODY: Loosely dubbed grey furon the waxed

THORAX: Olive floss or wool

HACKLE: Black hackle

INSTRUCTIONS: Lay a base of thread, then wrap the hook shank from bend to eye in lead wire. Tie in a length of silver tinsel chenille, then form a dubbing loop, paint it with sticky wax and lay some grey fur on the loop. Spin it and hold it out of the way while winding the tying thread, then chenille three-quarters of the way up the hook. Tie off the chenille and follow forward with the sparse, loosely dubbed fur. Tie off at the same point and add a length of olive floss or wool. Wind the thorax material forward, tapering it slightly towards the eye, tie off. Add a fine black hackle feather right behind the eye make two or three turns and tie off, whip finish and glue.

THE HELLDIVER LEECH

JULY 1986

Leech patterns are the ideal searching pattern for both experienced and inexperienced fly fishers alike. For the inexperienced, Leech patterns are near impossible to fish incorrectly. Fast, slow, deep and shallow all methods should work at any time. The best approach involves retrieves that simulate the rhythmic ribbon like swimming motion of a natural Leech. A slow to moderate paced hand twist retrieve works best when trout are not willing to chase. Use two- to four-inch strip pause retrieve on active fish and when covering water to locate concentrations of fish. When trout appear focused on other more difficult to imitate food sources such as Chironomids and *Chaoborus lava* (Glass Worms), play the greed card. A large seductive Leech stripped by the nose of a trout is a difficult morsel to ignore. Leeches are available in a wide range of colours and sizes. In clear lakes Leeches tend to be lighter in colour, with mottled black and brown varieties common. Darker and tannin stained waters produce darker species. Local observation is key, turn over rocks and logs along the shoreline as reclusive Leeches prefer these secluded hiding spots.

THE HELLDIVER LEECH

HOOK: #4 to #10 Mustad 9672

TAIL: Strands of fine burgundy mohair fibres mixed with black bear hair and black Marabou

UNDERBODY: Lead wire

BODY: Dubbed black and burgundy seal fur

WING: Same as tail, multiple layers tied Matuka style

INSTRUCTIONS: Start by putting a bend in the hook, upwards at a 30-degree angle about a third of the way down the shank. Attach the thread, lay a thin base and weight the forward section (ahead of the bend) with wraps of lead wire. Bind the lead down, tie off, coat with five-minute epoxy and set aside to dry. Re-attach the thread and tie in a long tail of burgundy mohair, black bear hair and black Marabou in a ratio of 50/50 black to burgundy. Start to dub the body, stopping every few wraps to tie in a segment of wing. Add five or six wing segments, stopping just short of the hook eye. Layer each segment so that it is just shorter than the previous one, and the first just shorter than the tail. Whip finish and glue.

CRYSTAL HAIR CHIRONOMID

MARCH 1988

AQUATIC INSECTS, SUCH AS CHIRONOMIDS and Mayflies, utilize air and gases, which they trap beneath their nymphal or pupal cuticle to aid the emergence process. The trapped air and gases aid not only their ascent but also the emergence process as well, as the gases help expel the adult from their aquatic medium. From a presentation perspective, the trapped air and gases provide a fish-attracting glimmer that is often the final decision factor for a prowling Rainbow. Successful imitations must incorporate this trait within their design. Ian Forbes' Crystal Hair Chironomid utilizes two strands of crystal hair or Krystal Flash for just this purpose. Anglers should keep in mind that too much of a good thing can alienate their pattern as bright and gaudy startles wary trout. The strategic addition of flash and sparkle allows the pattern to stand out in a crowd. During intense emergences added appeal is needed to draw a response. If trout are not focused upon size use a pattern on size larger than the naturals. When using strike indicators strip the fly using sporadic six- to 12-inch pulls. The strip raises the pattern and cascades it down once the pull is complete. Patrolling trout gravitate towards the pattern grabbing it the moment it returns to a suspended state. At times stripping the indicator is the only way of eliciting any response.

CRYSTAL HAIR CHIRONOMID

HOOK: #10 to #18 Mustad 94833 or equivalent 1X short shank

BODY: Dyed olive flank feather

RIBBING: Fine copper wire

THORAX: Dyed reddish-brown herl fluff

WING: Pearl crystal hair

GILLS (OPTIONAL): White emu or ostrich herl

INSTRUCTIONS: Fasten a short tail of olive fuzz, taken from the base of a dyed feather. Next, tie in a small strip of mallard flank feather and fine copper wire. Wrap the feather forward, leaving room for the thorax, and tie off. Follow with the wire ribbing, wrapping evenly in the opposite direction taken by the mallard flank. Tie in the crystal hair wing, only one strand, then fold it back and tie it down to form a "V" wing. Wind on a few strands of reddish-brown fuzz from the base of a feather for the thorax. If desired, a few wraps of white emu or ostrich herl can be added at the head to represent gills. Whip finish and glue.

CADDIS PUPA

MAY 1988

Once pupation is complete, the Caddis Pupa cuts its way free from its self made case home. After a few days prestaging along the bottom the Caddis is ready for the final ascent to transform into winged adult. During its transformation the Pupa developed elongated oar-like hind legs to propel it along its angled trek. An efficient swimmer, the Pupa sculls along using short choppy strokes, reminiscent of a Water Boatman or Backswimmer. After darting forward for a few short bursts the Pupa rests and then resumes its journey. The choppy swimming motion catches more than a few eyes and as trout become focused upon the Emerging Pupa they are quick to pounce, taking many Pupa as they pause. Fly fishers must mimic the Pupa's angled trek. Popular presentation techniques include weighted patterns in conjunction with a floating line, intermediate and stillwater lines. Make a long cast allowing the pattern to sink to the bottom. Begin a steady four- to six-inch strip retrieve. After four or five strips halt the retrieve, suggesting the natural breaks the Pupa takes during its angled trek to the surface. After a few seconds resume the retrieve until the cast is complete or the retrieve is stopped by the confident take of a Rainbow.

CADDIS PUPA

HOOK: #6 or #8 Mustad 94840 or equivalent

UNDERBODY (OPTIONAL): Fine lead wire

ABDOMEN: Pale-green antron dubbing, or dyed seal fur

OVERBODY: Pale-green antron

THORAX: Dyed rabbit fur, colour to match natural

LEGS: Dyed mallard, teal or guinea same colour as thorax

INSTRUCTIONS: Wrap the body with lead wire, this may or may not be necessary to sink the fly, depending on how much air the loosely dubbed body traps. Start the thread at the bend and tie in a long strand of dubbing material. Form a dubbing loop painted with sticky wax, tie in the abdomen fur, spin the loop and wrap it two-thirds of the way up the hook. Pull the tail over the abdomen as a loose veil and tie off. Use another dubbing loop to spin on the rabbit fur, ensuring it remains long and shaggy. Trim rabbit fur top and bottom. Tie in a few strands of mallard on either side of the body, whip finish and glue.

THE UNIVERSAL NYMPH

MARCH 1989

P RESENTATION, THE MANNER IN WHICH FLY fishers manipulate and place the fly in front of fish is the critical skill in fly-fishing, more so than pattern. When plying British Columbia's numerous productive stillwaters, this means choosing fly line and leader combinations to keep the pattern in the feeding zone and then selecting the correct retrieve for the conditions at hand. The retrieve is comprised of three components; the length the line is pulled, the speed of the pull and the over pace of these two actions. There are two main retrieves, the hand twist or hand weave and the strip. After making a cast, the rod tip should be lowered to, or in many instances placed, into the water. Many prey items dictate angler attention and there must be a direct connection between fly fisher and fly. Rod tips pointing skyward and inattentive provide an avenue of opportunity that feeding trout are only too willing to exploit. Place the fly line behind the forefinger of the rod hand once the cast is completed. Human hands and fingers are sensitive tools and are the first to sense a take. The fingers of the rod hand also provide the initial source of drag and in many instances are all that is needed to subdue trout.

THE UNIVERSAL NYMPH

HOOK: #4 to #18 Mustad 94840 or equivalent

TAIL: Speckled brown grouse body feather

BODY: Brown Marabou plume

RIBBING: Thin copper wire

THORAX: Brown Marabou and speckled olive partridge feather

HEAD: Peacock herl

INSTRUCTIONS: Start by tying in about 10 strands of grouse body feather at the tail, then a small clump of Marabou and a fine copper wire. Wind the plume three-quarters of the way up the hook, tie off but don't cut it. Follow with a few wraps of copper wire in the opposite direction, tie off and trim. Make three or four more wraps of Marabou to build a thorax, then add some partridge feather. For larger flies (#4 to #8) the whole feather should be wound on, for medium-sized flies (#10 to #14) only a few strands on either side are required and on the smallest flies (#16 and #18) this step can be omitted. The final stage is a head of peacock herl, wound on, tied off and varnished.

MARABOU LEECH

SEPTEMBER 1989

In recent years hanging small "Micro" Leeches beneath a strike indicator has become popular. When trout are cruising near or among aquatic and emergent vegetation the indicator method allows the fly fisher to present the pattern without hanging up. Imitating Leeches often results in familiar a phenomenon, short strikes. Short strikes are a result of three causes. If trout are not satisfied with a particular pattern or colour they often show interest but break off their charge at the last moment. Try changing the colour and size of the pattern to illicit a grab. Short strikes can also be caused by trout's approach to dining on Leeches. When threatened, Leeches often curl up in a defensive ball. A balled-up Leech is an easy mouthful for a trout, short strikes may be the result of trout trying to ball up the Leech pattern. When this happens cease the retrieve and wait for a strike as the pattern falls. Trout feed by flaring their gills, sucking their prey down their gullet. When a trout tries this approach on a Leech pattern nothing happens as the fly continues on its way stripped by the angler. If halting the retrieve fails resume stripping, in some instances the trout alters its attack striking the pattern broadside. Hang on tight, strikes are seldom delicate.

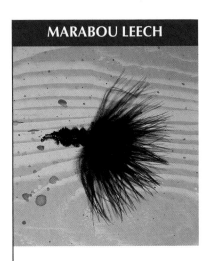

MARABOU LEECH

HOOK: #6 Mustad 9671 2X long shank
TAIL: Black Marabou
HEAD: Lead wire
BODY: Black Marabou, or chenille or fur or wool dubbing

INSTRUCTIONS: Start by making three or four wraps of lead wire around the shank at the head. Tie in a thick clump of Marabou, as long as the hook shank, for the tail. Filoplumes (after-shaft feathers) can be substituted as a tail material. If the Marabou is long enough, the remainder can be wrapped up the hook to the eye, creating the body. If the length of the Marabou is insufficient or filoplumes were used for the tail instead, then a body of chenille or dubbed wool or fur must be added in place of the Marabou. Whip finish and glue at the eye.

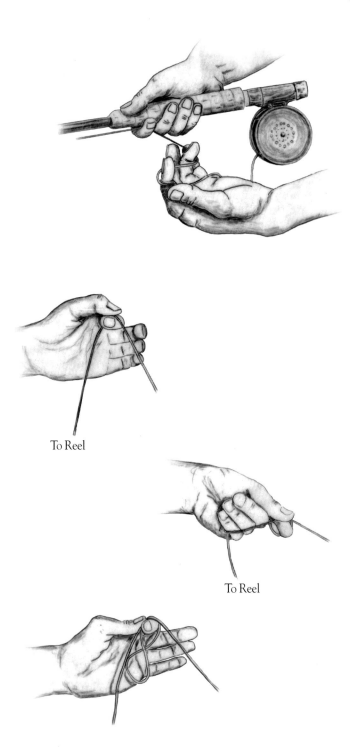

To Reel

To Reel

CHIRONOMID LARVA

MARCH 1990

Despite their frail feeble appearance, Chironomid Larvae are hardy creatures, capable of surviving in diverse environments. Many species create hemoglobin allowing them to survive in oxygen poor environments. This hemoglobin trait results in many species having a distinct red colouration and has lead to their common nickname of "Bloodworm." Hemoglobin can be increased at times to manage short-term oxygen deprivation such as the high heat of summer. As water temperatures increase its oxygen content decreases. Chironomid Larvae are ineffective swimmers. Removed from their tubular homes, Larvae move through the water column utilizing a vigorous head to tail lashing motion, a real eye catcher for foraging trout. Successful larval imitations often feature mobile materials to suggest movement. Popular materials include natural materials, such as pheasant rump fibres and Marabou. Synthetic fans should try Craft Fur or Anglers Choice's Shimmer. Adding the odd strip to the otherwise methodical retrieve further animates these materials. Chironomid Larva patterns work throughout the season. During the seasonal migrations of spring and fall Chironomid Larva often expose themselves to predation. During the spring, the Larva migrate from the depths to shallows and vice versa in the fall.

CHIRONOMID LARVA

HOOK: #8 to #12 Mustad 9479 5X short

TAIL: Dyed green, olive, brown and/or red Marabou or soft flue from the base of a body feather, colour to match naturals

UNDERBODY: Very fine lead wire

BODY: Green, olive, brown and/or red Frostbite mylar, colour to match naturals

INSTRUCTIONS: Tie in a Marabou tail in the appropriate colour(s) part way around the bend of the short shank hook. Vary the length of the tail to match the naturals you are trying to duplicate. Make a few turns of very fine lead wire around the hook shank near the head. Tie in a section of Frostbite and cover the shank and lead wraps from tail to head. Whip finish and glue.

BLACK LAKE SPECIAL

JUNE/JULY/AUGUST 1990

Just about any fly pattern benefits from a Peacock presence of some type. Peacock herl's only failing is its durability and to reduce pattern mortality some attempt at reinforcement is advised. For flies utilizing Peacock herl, a basic wire rib is a great starting point, but spinning the Peacock herl within a wire dubbing loop creates resilient Peacock bullets. Patterns capable of surviving a good chew. Begin by tying in the Peacock herl strands by the tips. Secure in a wire dubbing loop that is slightly shorter than the Peacock herl strands. The Peacock herl and the dubbing loop must be tied at the same point, at the rear of the hook after the tail has been tied in. Gather the Peacock herl pulling it parallel to one side of the wire loop. Using a pair of small push button electrical pliers grasp the herl and wire loop together at the bottom of the loop. Twist the wire and Peacock herl together a few times. Wind the loop forward a few wraps to start the body. Twist the remaining wire loop tight creating a durable Peacock noodle and complete the body. Starting the body in this manner avoids breaking the fragile tips of the herl during the initial phases of construction.

BLACK LAKE SPECIAL

HOOK: #8 Mustad 94840 or equivalent standard hook

BODY: Green mylar tinsel chenille (back third), peacock herl (middle third)

LEGS: 16 strands of dyed olive partridge or grouse body feather (eight on either side)

HEAD: Peacock herl

INSTRUCTIONS: Tie in a short piece of green mylar tinsel chenille at the bend of the hook, make three wraps and tie off. Pull back the fibres after each wrap to keep the fibres from being packed down by consecutive wraps. Tie in two long strands of good quality peacock herl with long fibres. Wrap the herl back and forth up the shank, building as much bulk as possible then tie off. Tie in eight strands of body feather on either side of the hook, then fasten two more peacock herls and wrap them ahead of the tying thread, back and forth to the eye. Tie off behind the herl head just ahead of the partridge legs. Whip finish and glue.

THE DRAGONFLY

MAY/JUNE 1991

Dragonflies hatch in late spring early summer by crawling along the bottom and out of the water. Trout plunder these plodding legions into the margins. Anchoring in the near shore shallows, cast out into deeper water and crawl and dart the fly back. Using a sinking line in conjunction with the countdown method allows the pattern to skim the weed tops on its migration toward shore. The countdown method is simple. Make a long cast and using a watch's second hand count the pattern down for a set time. Begin the retrieve. If the fly hangs up or fouls with weeds, reduce the sink time on the next cast by five seconds and so on until the fly fouls no longer. The goal is to time the pattern down, placing it at the level trout are foraging. After five or so casts fly fishers should have the level dialed in. Using the sweeping hand of a watch is more accurate than counting to oneself, especially when distractions and questions from others are present. The ideal fly should touch and grab weeds sporadically assuring the angler that the pattern is tracking at the correct level.

THE DRAGONFLY

HOOK: #4 3X long

ABDOMEN: Dubbed reddish-brown wool (upper portion) and tan wool (lower portion)

THORAX: Dubbed brown wool

LEGS: Brown, speckled grouse or partridge body feather

HEAD/EYES: Dubbed dark brown or black wool

INSTRUCTIONS: Cut up wool abdomen material into one-centimetre sections. Attach a strong tying thread to the bend of the hook and tie in the wool using two wraps in the centre of each piece so that the ends stick up and away from the hook shank. Tie a reddish brown piece on top of the hook, then a tan piece on the bottom in alternating order. Tie each section as closely as possible to the last until two-thirds of the hook is covered, then trim closely on the top and bottom and wider on the sides. Refasten tying thread and tie in a speckled feather and a short piece of brown wool. Wrap wool four times and follow over wool with feather wraps, tie off. Tie in the dark wool head using the same method as for the abdomen. This will represent the eyes and should be trimmed accordingly. Whip finish and glue.

BOB GILES' RIVER DRAGON

HOOK: #4 to #6 Mustad 9672 or equivalent long shank hook

THREAD: Olive

BODY: Dubbed olive natural or artificial sear fur

RIBBING: Thin silver tinsel

LEGS: Brown pheasant rump or pintail flank feather

HEAD: Deer hair and pheasant herl

INSTRUCTIONS: Using a dubbing loop started at the bend, wind on a pencil shaped body of seal fur. Tie in tinsel at the bend and make five or six evenly spaced wraps to the head and tie off. At the front shoulder of the dubbing tie in six fibres of the leg material of choice on either side of the body, trim forward butts. The legs should extend back to the point of the hook, and should flare out slightly from the sides of the fly. Tie the head of the river dragon in two parts. Start by laying a clump of deer hair at right angles to the hook and tie down with a series of figure-eight wraps. Tie in a peacock herl over the deer hair, spin it in the tying thread and wind several more figure-eight wraps over the deer hair. Whip finish and glue.

MO BRADLEY'S BLOODWORM

HOOK: #10 3X or #14 2X

TAIL: Black bear

BODY: Fine maroon chenille

Tie in a few barbules of pheasant rump at top and bottom.

INSTRUCTIONS: After wrapping an underbody of thread wind to bend and tie in a short stubby tail of black bear hair. Tie in fine chenille body material and wrap thread followed by chenille to within 0.3 centimetres of the hook eye. Tie off. Tie in a few barbules of pheasant rump top and bottom and whip finish. Roll the fly between thumb and forefinger and finish fly with head cement.

THE CRAYFISH

JULY/AUGUST 1991

CRAYFISH ARE A FAVOURED FOOD ITEM FOR not only trout but also other species such as Smallmouth Bass. Actually, Crayfish are the preferred food item for Smallmouth Bass. Many waters in British Columbia, including Lower Mainland, Vancouver Island and Sunshine Coast lakes, sustain large populations of these armor-plated Crustaceans. As with many invertebrates, Crayfish grow through a molting process. Recently molted Crayfish nicknamed "shredders" are a delicacy trout and bass seldom pass up. Shredder exoskeletons have yet to harden and are more palatable than a Crayfish in its normal case-hardened state. Fleeing Crayfish scoot through the water using rapid undulations of their abdomens and fan-shaped tails, pincers and antennae trailing behind. In this panic-stricken mode, Crayfish are vulnerable to the bold rush of an attacking bass or trout. More so than a cornered Crayfish in a defensive claw brandishing mode. Faced with a defensive Crayfish, bass and trout temper their assault waiting for the Crayfish to bolt before striking. Popular Crayfish haunts include rocks and wooden debris. These areas are not sympathetic to fly fisher or fly. Successful Crayfish patterns must be tied in an inverted style utilizing bead chain or heavy dumbbell eyes and tails secured down the bend of the hook.

THE CRAYFISH

HOOK: #4 or #6 2X long
THREAD: Waxed 3/0 nylon, monocord or Kevlar
EYES: Melted 80-pound test monofilament or bead-chain, painted black
CLAWS: Coarse reddish brown polyester wool, with epoxy tips
DUBBING: Reddish brown fur or wool
BODY: Amber Swannandaze
WHISKERS/UNDERBODY/TAIL: Dyed brown elk body hair, stacked to even ends
HACKLE (OPTIONAL): Brown hen hackle

THE CRAYFISH

JULY/AUGUST 1991

INSTRUCTIONS: Start by making the claws: Using a thick wool, brush out the strands so they are loose, then cover the tips with a fast drying epoxy. As the glue sets, flatten it with your fingers and hold until dry. When the glue is firm cut the shape of the claws with scissors. Tie in the elk hair whiskers and underbody, angling down. On top of the hook, opposite the point, tie on the eyes with figure-eight wraps and a drop of glue. Dub in a sparse section of fur or wool at the bend and wrap around the eyes, leave enough for two more wraps hanging. Tie in the claws on either side, behind and beneath the eyes. Ensure some soft wool is between the body and the epoxy part of the claw to allow freedom of movement. Trim the tip of Swannandaze and tie in. Wind the remaining dubbing on and tie off. Make two turns of a hackle feather, if desired. Wind the Swannandaze forward evenly to the eye and tie off behind the stub of elk hair that is left for a tail and tie it off. Seal head with glue, and trim hackle fibres, if you've used hackle, top and bottom.

THE RHYACO

WINTER 1991

CADDISFLY LARVA IS ONE OF THE MOST readily available food sources available to trout in rivers and streams. Free-living species such as the Rhyacophila or Green Rock Worm are a key dietary component for trout. Living amongst the bottom cobble and debris these predators are often swept from their perches. Large rivers such as the Thompson have impressive populations of Caddis including Green Rock Worms. First-time fly fishers plying large rives such as the Fraser and Thompson can be overwhelmed by both size and flow. The secret to these waters and other large rivers is breaking the river down into manageable chunks. Isolate riffles and appealing runs as though they were streams of their own. Narrow points of land are and small steam obstructions such as individual boulders and woody debris are other spots to investigate. Probe these near shore areas carefully as it surprising how often trout rest and feed around such obstacles. Too often anglers crash through, believing fish to be laying out in the deeper swifter flows. From an efficiency perspective, it doesn't make sense for trout to hold in heavy flows unless they have to. Fly fishers finding isolated runs, riffles, back eddies and side channels can spend an entire day searching these areas, enjoying spectacular results in the process.

THE RHYACO

HOOK: #4 or #6 Mustad 9479, down eye, 4X short

WEIGHT: Medium lead wire

THREAD: Yellow

UNDERBODY: Yellow floss

OVERBODY: Bright green Sealex artificial seal fur

LEGS AND TAIL: Dyed yellow speckled partridge

INSTRUCTIONS: Start by putting three or four wraps of lead wire near the head of the fly for weight. Wrap the lead wire and hook shank with yellow thread, leaving it at the bend. Tie in a short sparse tail of yellow partridge feather. Attach yellow floss at the base of the tail and wrap forward, completely covering the hook and lead wire then tie off at the head. Wind the tying thread back over the floss to the tail. Using a dubbing loop smeared with sticky wax, dub on a body of bright green artificial fur extending to the head of the fly. Add about 10 strands of dyed partridge for legs, whip finish and glue.

CALLIBAETIS

MARCH 1992

CALLIBAETIS IS THE DOMINANT MAYFLY species in B.C. lakes. Hatching as their name suggests in late May and through June, Callibaetis provide the first serious dry fly opportunity for stillwater fly fishers. Callibaetis Nymphs are active swimmers, moving through the water column by undulating their abdomens in rapid fashion. The Nymph settles for a moment fluttering down in a distinct arched profile, leg and tail splayed. The best retrieves mimic this behaviour and consist of a steady hand twist consisting of four or five rotations followed by a noticeable pause. Callibaetis favour the clear marl chara type lakes and Nymph patterns worked over the marbled shoals provides for exciting fishing. Prior to the hatch, Callibaetis Nymphs become restless rising and falling through the water column, a phenomenon known as false hatching. Seasoned fly fishers aware of this trait can extend their Callibaetis hunts for a few weeks prior to any hatch. Preferred presentation techniques involve floating, intermediate or clear tip lines and long leaders.

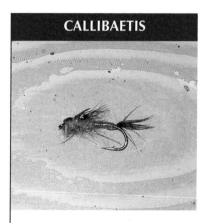

CALLIBAETIS

HOOK: #12 or #14 Mustad 94840 or equivalent

TAIL: Natural, speckled partridge feather

ABDOMEN: Sealex poly or rabbit fur dubbing, one-third each grey, medium-olive and tan

RIBBING: Fine copper wire

LEGS: Natural grey or tan speckled partridge body feather

THORAX/HEAD: Peacock herl

INSTRUCTIONS: Blend the three colours of dubbing material in a coffee grinder to mix it evenly. Tie in a tail consisting of five or six strands of speckled grey partridge feather at the bend of the hook, extending back about half the total hook length. Tie in a small section of copper wire at the base of the tail. Using a dubbing loop and a small amount of body material, wind the dubbing two-thirds of the way up the hook and tie off. Cover the dubbing with even wraps of the wire ribbing in the opposite direction of the body wraps. Tie off and trim excess wire. At this point tie in two strands of well feathered peacock herl, make two wraps and leave the herl hanging (on hackle pliers). Tie in six or seven strands of speckled partridge to imitate legs, they should not be too long, about half the hook length. Continue winding the peacock herl forward to the eye ahead of the tying thread, then back over itself to the tying thread. Tie off, whip finish behind the peacock head and add a drop of glue.

Rain is no deterrent...

PLASTIC CHENILLE SHRIMP

HOOK: Tiemco #3761 #10 to #12 2X long shank or equivalent

TAIL: Olive-green hackle fibres, tail length five millimetres

BODY: Olive or dark green plastic chenille

INSTRUCTIONS: Tie in 15 to 20 hackle fibres at the bend of the hook for the tail. Tie in plastic chenille and wind tightly to eye of the hook and tie off. Trim plastic chenille fibres from back and side of the fly to create the shrimp outline. Take a match or lighter and pass the flame close to the fly to melt down the plastic and form a smooth back.

FROSTBITE BLOODWORM

HOOK: #10 to #16 Partridge GRS15ST Klinkhamar Special

THREAD: Red Gudebrod 8/0

TAIL: One strand of Super Floss, split

RIBBING: Fine gold, copper, silver or red wire

BODY: Red, green or maroon Frostbite

HACKLE: Red Angel Hair (optional)

INSTRUCTIONS: Start by tying in a short piece of Super Floss, split into individual strands, even with the hook point. Tie in wire ribbing above tail, followed by Frostbite body material. Wind thread, then Frostbite to the hook eye, wrap securely. Wind the ribbing wire in six to eight evenly spaced wraps to the head and secure. Tie in a hackle of Angel Hair, extending the length of the hook if desired. Whip finish and glue. Add a coat of brushable super glue to this (and any other chironomid) pattern for added shine and durability.

BAGGY SHRIMP

APRIL 1993

SCUDS, AS WITH MANY TROUT FOOD SOURCES inhabit areas that prove challenging to probe with the fly. Preferring the safety of weed beds, marl and rocks, Scuds provide unique presentation challenges. Deer hair and sheet foam provide the innovative fly tier the flexibility to create buoyant patterns. Fast sinking fly lines that would normally overpower the presentation now prove valuable. The sink rate of the line drags the floating pattern down to the foraging trout. The pattern bobs above the obstructions and debris permitting the slow random retrieve needed to simulate a foraging Scud. Leader length needs to be short to keep the offering skimming the weed tops. Four feet or less is adequate. Trout feeding over marl flats are skittish. Using a buoyant pattern on a sinking line is ideal. The fly line disappears into the marl eliminating the shadows a floating or slow sinking line might cause. Cast once the line has settled give it a quick strip to create a convincing puff of marl drawing trout to the fly.

BAGGY SHRIMP

HOOK: Mustad 9672 or similar
BODY: Pale olive dubbing
RIB: Four-pound-test monofilament
UNDERWING: Deer hair
WINGS: Decorating plastic

INSTRUCTIONS: Start the fly by wrapping an appropriate colour tying thread to the bend of the hook. Next, fasten a short length of fine monofilament line, (four-pound-test) which will be used as a rib. Using a dubbing loop smeared with a sticky wax (cross country ski klister), spread on a layer of dyed olive dubbing. The dubbing can be natural fur or a synthetic. Wind the dubbing to near the end of the hook and tie it off. Next, comes the underwing of deer hair. Even the tips of the hair with a stacker and fasten a small amount on the top of the hook. The hair should extend just a short distance beyond the end of the hook. Then fold a thin strip of decorating plastic over the hair and tie off near the head. The plastic should cover the hair but not the fur body. When the plastic is in place, wind the monofilament rib over it with four wraps. It is necessary to hold the deer hair and plastic wing in place as you wind on the rib. Other wise, the wing will spin around the body. Tie off the rib, whip finish and glue. To complete the fly, pluck out some dubbing with a needle. This imitates the shrimp's legs. On larger patterns, a few strands of speckled partridge can be added near the eye of the hook.

MARCH BROWN BEAD HEAD

MAY 1993

TUMBLING NYMPH PATTERNS ALONG THE bottom of rivers and streams is arguably the best tactic for consistent success. In the absence of a hatch, trout position themselves in lies offering protection from current and predators, comfort and proximity to food. Constant current surges dislodge countless organisms trout are only too willing to prey upon. Many aquatic insects partake in a phenomenon known as behavioural drift. At certain times of the day, Mayfly Nymphs, Stonefly Nymphs and Caddis Larva release their benthic grip dispersing downstream. Behavioural drift ensures adequate population densities throughout a system. Some species disperse due to their own development. Early in their growth a specific species may favour the slower calmer sections of a river. As the Nymph or Larva develops, its living conditions change dictating a move to such as riffles or pools. Fly fishers bouncing Nymphs along the bottom provide a realistic presentation, as trout are accustomed to lunch tumbling towards them and take Nymph, Larva and Pupa patterns without hesitation. The best patterns are simple scruffy affairs using an "in the round" style that provides a consistent profile for trout no matter how the current stumbles and bounces the fly along.

MARCH BROWN BEAD HEAD

HOOK: No. 10 to 14
DUBBING: Muskrat, mink or rabbit
RIB: Yellow floss
WING: Mottled grouse feather
BODY: Partridge feather

MARCH BROWN
BEAD HEAD

MAY 1993

INSTRUCTIONS: Start by pinching the barbs on the hooks to be used. Depending on the thickness of the hook wire, it might be necessary to make a few turns of thread at the eye of the hook. Tie it off and add a drop of glue. Before the glue sets, slide the metal bead over the point and up the shank to the hook's eye. Start tying the thread behind the metal bead to hold it in place, then wind the thread to the bend of the hook. Start the fly by tying in a tail, using six strands of partridge or grouse body feather. The tail should be half the hook length. At this point, tie in a short length of yellow floss that will be used as a rib. Using a dubbing loop dressed with a sticky wax (cross-country ski klister), spread on a thin layer of muskrat, mink or rabbit's hair. The hair should be tapered so it will get progressively thicker near the front of the hook. The colour should be a mixture of grey and brown. Leave the guard hairs in to give a shaggy appearance. Wind the dubbing to just behind the metal bead and tie it off. Next, evenly space about four wraps of the yellow floss over the fur, in the opposite direction to the way you wound the dubbing. Tie off the floss and add as short wing of mottled grouse feather. Remove the fuzz from the base of a well-marked partridge feather and tie the feather just behind the metal bead. Holding the feather by its tip, make one or two wraps and tie it off behind the metal bead. Whip finish and glue.

THE WIGGLE FLY

JANUARY/FEBRUARY 1994

As a predator, trout must be aggressive to survive. The trout's aggressive tendencies can be manipulated to the fly fisher's advantage. When trout are not responding to mainstream presentations this approach is often the only method that provides any kind of consistency. Immediately after ice-off, trout mill around still under the effects of their winter stupor. With little or no insect activity to shake them from their trance, fly fishers must choose active presentations and mobile fly patterns to throw the odds in their favour. This approach also works when trout focus upon the minute. Zooplankton and Chaoborus feeders need to be shaken from their routines. Patterns featuring movement and seductive action are the ideal choices for these conditions.

THE WIGGLE FLY

HOOK: 9672 Mustad #6 or #4

TAIL: Dyed aftershaft body feather or marabou

HACKLE: Dyed partridge speckled body feather

WIGGLE HEAD: Friendly Plastic molded into disc-shape, with clip or tied with loop knot.
(Note: Friendly Plastic is moldable hard plastic, usually found at a hobby shop. It softens in hot water and can be shaped with your fingers.)

INSTRUCTIONS: In a fry pan, heat some water to just below the boiling point. Immerse a stick of Friendly Plastic and let soften. Pinch off tiny pieces and form them into thin discs about half the size of your pinky fingernail. Using another tiny piece of Friendly Plastic, fasten the disc to the kink on the hook. The disc needs to hang down below the eye of the hook and should have a slight, cup-shaped face. At this point, paint the Friendly Plastic an appropriate colour of an insect, or paint on eyes. Behind the plastic head, wrap the tying thread to the bend of the hook. Then, using two or three filoplumes, add a tail of about the same length as the hook. The remainder of the filoplume is wound around the hook to form part of the body. Continue the body by winding on other filoplumes. Warp this to just behind the plastic head. Add a hackle of well marked partridge feather. Only two wraps are necessary. Whip finish and glue to complete the fly.

GOLDEN STONEFLY

MAY 1994

T HE STONEFLY NYMPH IS A RIVER AND stream staple. Every body of flowing water containing oxygenated water found in riffles and pocket water is home to Stoneflies. Stonefly Nymph possess a rudimentary gill system requiring oxygen rich water to be forced by them. Place a Stonefly Nymph in a dish and within seconds it begins a series of push ups in an attempt to breathe. Stoneflies are feeble swimmers and when swept from the bottom cobble they tumble and roll until regaining their footing or disappearing down the throat of an opportunistic trout. One of the most common species found in British Columbia rivers and streams is the Golden Stone. This large Nymph is an active predator scouring the bottom substrate for Mayfly Nymphs, Caddis Larva and other forms of fodder. This prowling trait results in Golden Stone Nymphs being constantly swept from the rocks. Golden Stone imitations need to be rolled along the bottom to suggest a current swept natural. As a result these patterns need to be weighted, often to excess to consistently place the pattern near the bottom where waiting trout find respite from the swift flows above. This mandatory requirement for success spells a high casualty rate on the fly box as many Stonefly Nymph patterns become lodged amongst the rocks.

GOLDEN STONEFLY

HOOK: #6 or #4 2X to 3X long

BASE: Thick lead wire (rear half) packing foam (front half)

TAIL: Two dyed yellow goose biots (short, thick strand of wing feather)

UNDERBODY: Yellow floss

BODY: Flat monofilament

WINGS: Wide strip of lacquered brown pheasant tail

THORAX/HEAD: Dubbed amber yellow rabbit fur

LEGS: Rubber (Optional)

GOLDEN STONEFLY

MAY 1994

INSTRUCTIONS: Build up a thread underbody and fasten two short pieces of thick lead wire to either side of the rear half of the hook. Wrap to hold it and glue lead securely. Wind a strip of packing foam on the forward half of the hook and wrap down the shank, leaving the thread hanging at the bend. Tie two short thick goose biots for a tail, followed by a strand of yellow floss and a length of flat monofilament. Wind thread halfway up the hook, and follow with a smoothly tapered body of yellow floss. Wind the mono on top of the floss in tight, even loops, tie it all off and trim. Place a wide, lacquered strip of pheasant tail at the bend and tie on. Make a dubbing loop, sticky wax it and spin on rabbit fur. Make two wraps, fold pheasant wing forward and tie it down. Repeat the last step to make a second wing case. Let pheasant tips remain and make a head with two more turns of the dubbing. Whip finish and glue. Trim fur around the head and from the underside, but allow it to stick out on the sides to simulate legs. Trim out the centre of the pheasant tip, leaving a strand on either side to form the antennae. On larger flies, thread rubber legs through the body with a needle and fasten with glue.

GOLD BEAD MARABOU LEECH (BLACK)

HOOK: #12-#8 Tiemco 5262 2X long shank

HEAD: Umpqua Feather Merchantt gold bead (x-small for #12, small for #10 and #8)

THREAD: Black

TAIL: Black Marabou fibres

BODY: Same as tail

RIBBING: gold wire

INSTRUCTIONS: Pinch down barb and slide on gold bead to the eye of the hook. Attach thread behind bead and build up a shoulder to secure bead in place. Elect a 1.5 centimetre wide group of fibres from a long marabou feather, remove from centre quill and tie in at tail no more than 1.5 centimetres longer than the hook shank. Tie in gold wire at the bend. Grasp the marabou by the butts and twist together about six turns, then wrap the spun marabou forward up the shank towards the bead adding and spinning more fibres as necessary to reach the head. Tie off. Wrap the gold wire forward to bead in the opposite direction of the marabou wraps, tie off.

SELF CAREY

HOOK: #8 to #12 Mustad 94833

BODY: Three ringneck pheasant tail fibres, wrapped

RIBBING: Black thread or fine gold tinsel

HACKLE: Very small brown ringneck pheasant rump

BLACK MICRO LEECH

HOOK: #8 to #12 TMC 2457

HEAD: Silver bead

UNDERBODY: Fine lead wire

TAIL: Black rabbit strips and a few strands of black holographic Flashabou

BODY: Black rabbit steel head black dubbing (has a green flash)

RIBBING: Silver wire, red wire (double ribbed)

Nymphs 43

The days of rubber waders.

ERNIE'S SCUD

MAY 1995

Trout have a penchant for dining on small things, Chironomids, Zooplankton and Scuds — Hyallela to be specific. During the fall feeding binge stillwater trout embark Hyallela are a prominent component. Most stillwater fly fishers shudder at the thought of fishing a tiny size 14 or smaller Scud patterns. Yet most anglers wouldn't blink an eye at tying on a similar sized Chironomid pattern in the spring. Small patterns do not need to be complicated. When it comes to Hyallela impostors, simple dressings such as Ernie's Scud are perfect. Hyallela are typically a pale olive or almost tan colouration depending upon their surroundings. Anglers may observe blue Scuds, identifying that a recent molt has occurred. With their exoskeleton not fully hardened, the copper ion within the Scud's body fluids create a distinct blue hue. Wise fly fishers always keep a stash of a few blue Scuds just in case. In recent years curved scud hooks have become increasingly popular. These style hooks are ideal as they sink and hook well. The curved posture they create does an excellent job duplicating the semi-curved posture swimming Scuds display. Too much curvature detracts from success. Comma shaped Scuds are dead, resting or feeding. Fly tiers need to be aware when weighting curved hooks. Once weighted, the high side of the bend is now heaviest point and causes the fly to roll upside down due to the keel effect. Weighted Scud patterns should be fished using straight shank hooks.

ERNIE'S SCUD

HOOK: #8-#16 round bend

BACK/TAIL: Three strands of Pearlescent Krystal Flash

BODY: Blue, green, olive, grey or orange dubbed wool or synthetic

RIBBING: Fine copper wire (optional)

INSTRUCTIONS: Tie in three strands of Krystal Flash at the hook bend, leaving a short tail. Tie in wire at this point if a slow sinking fly is desired. Using a dubbing loop and the appropriate coloured wool or synthetic dub on the body, leaving it quite shaggy. If tying in a wire ribbing, wind it in four or five wraps to the hook eye and tie off. Fold the Krystal Flash back over and tie in behind the hook eye, leaving a short stub at the head. Whip finish.

KEVIN'S ORANGE SHRIMP

JUNE 1995

Scuds are the peak stillwater staple providing a solid backbone for quality trout fishing. Rich in protein, Scuds enable trout to pack on pounds at an alarming rate. Growth rates of two pounds within a single season are common. The calcium rich lakes of south-central British Columbia provide ideal habit for both families of Scud, Gammarus and the small Hyallela. Gammarus require calcium to facilitate exoskeleton development and are not as widespread as Hyallela. Prolific breeders, Scud populations often reach epidemic proportions. Pregnant females store their eggs in an abdominal cavity called the marsupium. The pending prodigy gives the pregnant females a distinct orange hue drawing the attention of trout. Focused feeding upon pregnant females is common and proves frustrating to the inexperienced. Scuds are best imitated using a slow sinking intermediate or clear stillwater line. Keep the retrieves varied matching the Scuds erratic swimming nature. A six- to 12-inch draw followed by a noticeable pause works best. Probe weed beds and rocky margins for best results, especially after a good breeze has pounded the lake.

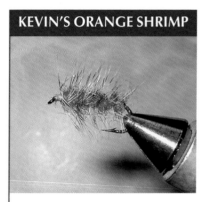

KEVIN'S ORANGE SHRIMP

HOOK: #12 Mustad 94840 or equivalent
THREAD: Pre-waxed brown
BODY: Orange chenille
BACK: Reddish-brown pheasant rump
HACKLE: Well-marked grizzly, palmered

INSTRUCTIONS: Moisten the rump feather and roll it between your fingers forming a narrow clump. Tie it in near its tip using about five wraps and leaving about seven millimetres for a tail. Strip a small section from the tip of the chenille and fasten it just in front of the feather. Then tie in a grizzly hackle by its tip in the same spot. Stroke the barbules tip to butt to make them separate and stand out. Wind on the chenille body with tight, even wraps, making the first wrap right under the pheasant feather. Tie off three millimetres back of eye and trim any excess. Palmer the hackle with five wraps over the chenille and tie it off. Fold pheasant rump feather back over and tie it down at the eye. Trim excess, make a neat head, whip finish and glue.

FRESHWATER SNAILS (FLOATING AND SINKING)

NOVEMBER/DECEMBER 1995

Although Snails are found in a wide range of sizes, trout can only feed upon Snails size 12 and smaller. It is a simple function of digestion as trout cannot ingest or pass large Snails. Some interior trout lakes are notorious for Snail feeding trout, much to the fly fisher's chagrin. Snail feeding often occurs in late summer and can carry on into the fall. One of the challenges facing the fly fisher is determining if trout are preying upon Snails with any regularity. Trout foraging upon Snails on or near the bottom do so by moving through the underwater jungle grabbing dislodged Snails and plucking them from aquatic vegetation. Clogged climes such as weed beds and woody debris are tough to fish. Strike indicators allow fly fishers the opportunity to suspend a sunken Snail offering just above the weed tops at trout level. As with other strike indicator presentations stripping the pattern from time to time might be needed to draw attention to the pattern.

FRESHWATER SNAILS (FLOATING AND SINKING)

HOOK: #14 English bait hook or 4X short shank

UNDERBODY: Lead wire

BACK: Two long strands of peacock

BODY: Spun deer hair (floating), bronze peacock herl and fine copper wire (sinking)

HACKLE: Filoplume feather (sinking)

INSTRUCTIONS:

Floating: make two wraps of lead wire around the centre of the hook. Fasten two long strands of peacock at bend, and spin a body of deer hair along the entire hook. Tie off the body and trim it very close on the back, leaving it longer on the inside bend. Re-attach the thread near the eye and wind the peacock in even wraps along the deer hair body. Tie off, whip finish and glue.

Sinking: Make three turns of lead wire at the hook's centre, then wrap thread well around the hook's bend. Fasten a length of fine copper wire, at least four long strands of peacock and a small filoplume feather. Wind filoplume three times like a hackle, tie off. Twist peacock strands around the copper wire and wrap the whole clump around the hook to make a thick body, adding peacock strands as necessary. Whip finish and glue.

FLASHABOU SCUD

MAY 1996

Scuds or freshwater Shrimp spend the majority of their time rummaging through and under dense weed mats, rocks and other sunken debris. Instinctively reclusive Scuds seldom venture out into open water especially during the height of the day. However, under the low light conditions of morning, evening or overcast days Scuds venture out to feed. This active cycle often coincides with that of foraging trout. Prior to the start of any hatch activity fly fishers should start their day probing likely areas with Scud patterns. Patterns with a hint of sparkle or flash work are ideal as they reflect any available light to attract fish. Using light as a form of protection, trout patrol the shallows dining the numerous darting scuds. Powered by a steady hand twist retrieve a Scud pattern in conjunction with a 15-foot or longer leader is an excellent strategy. Allow the pattern time to sink before commencing the retrieve. The Scud pattern should skim the weed tops as the naturals do. In most stillwater situations depth is the critical factor. Trout for the most part are selective upon depth and opportunistic on food source.

FLASHABOU SCUD

HOOK: #10 to #16 short shank
RIBBING: Fine copper wire
BACK: Bronze peacock herl
BODY: Dubbed gold Flashabou

INSTRUCTIONS: Wrap thread to bend and tie in a strand of fine diameter copper wire. Over the wire tie in eight strands of bronze peacock herl. Form a dubbing loop and paint it with sticky wax. Prepare a mixture of green and gold Flashabou into five millimetre long strips and spin it into a tight chenille. Wrap dubbed body up to the hook eye and tie it off with at least five wraps of thread. Pull the strands of peacock back over the body and wrap in at hook eye, trimming any excess. Wind the copper wire in four of five even wraps, ensuring that the wire does not bind down the Flashabou body by picking out with a fine tweezers. Whip finish and trim Flashabou neatly.

SHAGGY DRAGON

JULY/AUGUST 1996

Dragon Nymphs are a large, calorie rich food source. With multiple year life cycles, Dragon Nymphs are available year round. Trout are accustomed to seeing Dragonfly Nymphs and they are a welcome addition to the trout's palette. Presented properly, Dragon Nymph patterns are seldom refused. Capable of absorbing water through their abdomens and expelling it through their posterior, all Dragon Nymphs come afterburner equipped. Masters of camouflage and stealth, Dragon Nymphs prefer a more sedentary approach utilizing their internal propulsion system to avoid a threat or for a final bull rush on a prey item. Dragon Nymphs are best duplicated using slow sinking lines such as the intermediate or stillwater. The retarded sink rate of these lines allows for a slow, plodding retrieve without the risk of the fly line over powering the presentation. Use a slow steady hand twist retrieve. Mix in the odd four-inch strip to suggest a jetting Nymph. Place strategic pauses within the retrieve keeping the presentation varied, reminiscent of a patrolling Nymph. If trout are in the mood to chase, faster sinking lines in conjunction with a brisk four- to six-inch strip retrieve to suggest darting Nymphs.

SHAGGY DRAGON

HOOK: Mustad model 9671 hook #4 or equivalent

THREAD: Danville's waxed, olive-coloured thread

EYES: Bead chain or closed cell foam (punched from a beach sandal)

TAIL: Dyed, mottled chicken body feather, and two clumps of dyed Marabou body feathers

BODY: Dyed chicken filoplumes (after-shaft feathers)

LEGS: Dyed guinea fowl hackle

HEAD: Marabou

INSTRUCTIONS: Tie in eyes three millimetres back from the eye of the hook using a series of figure eight knots, and secure with a drop of glue to the thread. Wind thread to the bend and tie in a mottled chicken body feather, of about the same length as the hook. Tie in a clump of Marabou on both sides of the tail. To form the body wrap a series of filoplumes like hackle right to the eyes (soft fluff from base of body feathers tied with a dubbing loop and sticky wax can be substituted for filoplumes). Tie in legs behind eyes. Wind a bit more Marabou around the eyes and tie off at the hook eye, whip finish and glue.

STELLAKO TUMBLING STONE (ADAPTED)

HOOK: #6 to #12 2XL
UNDERBODY: Lead wire
HEAD: Gold bead
TAILS: Brown goose biots
BODY: Brown dubbed wool, with some flash particles mixed in
BACK/WINGCASE: Brown Raffia (Swiss Straw), coated with clear nail polish
THORAX: Ostrich herl
RIBBING: Brass wire
LEGS: One grizzly and one ginger hackle (clipped top and bottom)

ARIZONA DRAGON

HOOK: #4 2XL
THREAD: Black, brown or dark green
TAIL: Pheasant tail
UNDERBODY: Wool or chenille
OVER BODY: Arizona synthetic peacock spun in a thread dubbing loop, and further spun with a fine green wire-dubbing loop
LEGS: Knotted pheasant tail, six strands for each leg
THORAX: Four or five strands of peacock herl spun with a fine green wire-dubbing loop
WINGCASE: Olive Raffia, Swiss straw, Scud-Back or Rainy's Stretch-Flex
EYES: Olive or black ultra chenille, knotted

PEACOCK WATER BOATMAN

SEPTEMBER/OCTOBER 1996

Distant cousins Water Boatman and Backswimmers are often confused and intertwined by anglers.

Both insects breathe air and must constantly return to the surface to grab a bubble of air known as a plastron for respiration. As a result these insects tend to be prisoners of the shallows although it is not unheard of to have intense activity over deeper water. The air bubble creates a distinct silver sheen to the insect obscuring their natural colouration in many instances and is an important pattern trait. Water Boatman reach a maximum size of three-eighths of an inch and are characterized by dark backs and light bellies, best described as the colour of masking tape. Backswimmers are larger than their vegetarian cousins and feature light backs and dark bellies. Backswimmers are entirely predacious and one of their favourite foods are Water Boatman. Handle Backswimmers with caution as they can inflict a nasty bite. Hook sizes for Water Boatman range from size 10 through 14. Backswimmers are best imitated using size eight and 10 hooks. Both insects feature prominent oar-like hind legs that are key components to imitate. Popular leg materials include Super Floss and Sili Legs.

PEACOCK WATER BOATMAN

HOOK: #10 or #12 short shank round bend
UNDERBODY: Fine lead wire
LEGS: Grey rubber leg material
TAIL: Silver Flashabou
WING CASE: Eight to 10 strands of bronze peacock herl
BODY: Yellow Flashabou

INSTRUCTIONS: Build up a base of tying thread on hook and wind lead wire from mid-point to the eye. Fasten a one-inch section of leg material over the lead wire at the mid-point with figure-eight wraps. Wind thread to the bend and tie in a short tail of silver Flashabou. Attach peacock herl on top of the tail. Dub Flashabou on waxed thread, wind forward to the hook eye and tie off. Pull wing case over the dubbing and tie off behind the eye. Leaving a short head extending beyond the eye. Whip finish and glue. For protection, cover wing with a thin layer of silicone glue.

DEER WING BACKSWIMMER		SEPTEMBER/OCTOBER 1996

DEER WING BACKSWIMMER

HOOK: #10 or #12 short shank round bend
UNDERBODY: Fine lead wire
LEGS: Grey rubber leg material
TAIL: Silver Flashabou
WING CASE: Light deer hair
BODY: A mixture of gold and green Flashabou

INSTRUCTIONS: Build up a base of tying thread on hook and wind lead wire from mid-point to the eye. Fasten a one-inch section of leg material over the lead wire at the mid-point with figure-eight wraps. Wind thread to the bend and tie in a short tail of silver Flashabou. Fasten the deer hair wing case at the bend. To form the body cut several four millimetre lengths of Flashabou and spin it on a dubbing loop after applying a sticky wax, then wind on to the eye of the hook and tie off. Pull wing case over the dubbing and tie off behind the eye. Leaving a short head extending beyond the eye. Whip finish and glue. Use a black felt pen to mark the wing case and head with narrow bands, then cover wing with a thin layer of silicone glue to protect it.

BEAD-HEAD MICRO-MINI LEECH

JANUARY/FEBRUARY 1997

Since their initial introduction from Europe bead-head patterns have exploded onto the North American fly pattern scene. Just about every pattern seems to have beads blended into their designs. Fly tiers have many choices, metal tungsten and glass plus a cornucopia of colour. Popular metal styles in metal and tungsten include black, copper, gold, silver and white. White metal and glass beads are popular with Chironomid tiers. Bead sizes are often questioned. For sizes eight and 10 go with one-eighth-inch, sizes 12 and 14 go with 5/32-inch and for sizes 16 and smaller opt for 5/64-inch beads. The goal is to provide a proportional balance of weight and colour. Glass beads are popular pattern additions not so much for the weight but rather for the seductive iridescence and translucence they provide. Depending upon how the pattern is presented, it is possible to have one pattern tied in glass, metal and tungsten modes. Glass bead versions for slow shallow presentations, tungsten beads for a quick descent patterns to explore deep water or fast flows. Don't limit the use of beads to the front of a pattern either as beads can be planted anywhere within the pattern suggesting features such as egg sags on diving Caddis patterns and air bubbles on Water Boatman.

BEAD-HEAD MICRO-MINI LEECH

HOOK: Captain Hamilton Partridge #16 or short shank, round bend equivalent

HEAD: Glass Bead, blood-red in colour

TAIL: About six strands of black, brown, or olive Marabou

BODY: Same as body

INSTRUCTIONS: Pinch down the barb with a pair of fine needlenose pliers and set hook in vice. Build an underbody of thread on hook shank and slide the bead on to the hook. Position bead right behind the eye and glue to set in position. Refasten thread near tail and tie in Marabou with tips extending back as a short tail. Wind the remainder of the Marabou around the shank to behind the bead-head and tie off. Whip finish behind the bead eye.

BLUE-WINGED OLIVE

MAY 1997

Blue Winged Olive are the most widespread species of Mayfly in Western North America. These diminutive Nymphs from the Baetidae family should be in every serious river and stream angler's fly box. At the beginning of the season these hardy Nymphs will be at their largest and best imitated with size 14 or 16 imitations. As the season progresses and successive hatches occur overall size reduces due to the limited growth cycle between generations. By the end of the season fly fishers working the prehatch period using Nymph patterns must now use size 18 or smaller Nymphs. These tiny sizes may cause some anglers to shudder but trout have no problem honing on these swimming Nymphs. Blue Winged Olive or Baetis Nymphs live in diverse habitats and substrate, capable of surviving in runs, glides and riffles. Imitating Baetis Nymphs is straightforward using a floating line, nine- to 12-foot leader and strike indicator. Once the fly has completed the drag-free portion the presentation allow the pattern to swing to the shore. Trout following the pattern are attracted by the elevated swing as it simulates escaping food and the emergence rise of a Nymph.

BLUE-WINGED OLIVE

HOOK: #14 to #28 1X short shank
THREAD: 6/0 olive, acts as rib as well
TAIL: Four olive dyed moose mane hairs
BODY: Moose mane
HEAD: Dubbed dark grey wool
THROAT: Dun dry fly hackle

INSTRUCTIONS: Fasten moose mane hairs at bend of hook, tails extending slightly longer than one hook length beyond bend. Wind the remainder of the moose hair to halfway up the shank. The wrap tying thread three or four times spaced evenly to create the rib, tie off and trim the ends. Use a dubbing loop to twist in a small ball of wool close to the hook shank. Fasten a dun hackle feather by the butt and wind on, keeping hackle close together. With moistened fingers, pull the hackle below the hook shank, flip the dubbed wool over top of the hackle and make a half turn at the head. The wool will hold the hackle below the hook shank if tied in properly. Tie off the dubbing loop, make a nice head, whip finish and glue head to complete the Blue-Winged Olive.

LOHR'S DOWN AND DIRTY DRAGON

MAY 1997

IN ORDER TO BE SUCCESSFUL IMITATING Dragons fly fishers must be able to explore weedy unfriendly territory. Traditional weighted or less than buoyant patterns are at constant risk of fouling. Patterns incorporating variegated colours of spun and clipped Deer hair are popular with serious Dragon Nymph impersonators. Within the Dragonfly clan there are two families. The slender hourglass-shaped Darners and the squat Spider-like Sprawlers. Most are familiar with Darners as their bold aggressive nature keeps them in the limelight more often than their shy reclusive cousins. Despite their timid nature Sprawlers are widespread and are a favoured trout tidbit. On the clear chara and marl lakes of the south-central interior sprawlers inhabit chara beds in alarming numbers. Smothering themselves with weed growth and debris lashing out whenever and unsuspecting Mayfly Nymph, Scud or other unlucky food items stumbles by. Although Sprawlers contain the same afterburner system as Darners, they seldom use it. Preferring to plod along the bottom with a disposition matching their introverted lifestyle. Presentation must match the natural Nymphs to be successful, hence the need for buoyant patterns. Use a slow methodical hand twist retrieve and be aware for sometimes gentle takes as the trout plucks the pattern from its path.

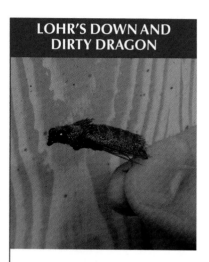

LOHR'S DOWN AND DIRTY DRAGON

HOOK: Daiichi 2220 #6-#8
THREAD: 8/0 black Uni-Cord
BODY: Olive and brown spun deer hair, medium to coarse
LEGS: Ringneck pheasant tail fibres
EYES: Three millimetre black plastic bead chain
HEAD: Olive and brown spun deer hair

INSTRUCTIONS: Tie in and clip body into an oval or egg shape. Tie in a few pheasant tail fibres on each side to represent the legs. Cross wrap bead eyes just back of hook eye. Tie in head and trim to shape.

FLOATING SNAIL

MAY 1999

Despite angler protests Snails often end up on the trout's dinner plate. Angler angst aside trout can be taken using Snail patterns. Snails often let go of their bottom hugging grip drifting up to the surface, using their foot to slide along the underside of the meniscus. If the quantity is concentrated trout can and do dine on bobbing Snails. At first glance, rise forms to surface Snail feeders are reminiscent of an adult Chironomid rise. Throwing the usual Chironomid fare proves frustrating. Should an angler manage to dupe a fish its hardened maraca feel is a dead giveaway that Snails are on the menu. Floating Snail patterns need to be constructed so that they bob in the surface film. Ian Forbes' Floating Snail incorporates a balance of lead wire and spun and clipped deer hair to achieve this trait. Wind drifting a Floating Snail through a pod of surface feeders is an ideal strategy. Make note of where the pattern lands and track this area throughout the drift. Any rise should be met with a smooth raising of the rod. Let the startled reaction of the fish do the rest.

FLOATING SNAIL

HOOK: Tiemco 100 or similar fine-wire, dry fly hook
UNDERBODY: Ethafoam strip
BODY: Peacock herl
RIBBING: Copper Krystal Flash
HACKLE: Brown hackle

INSTRUCTIONS: Tie in a thin ethafoam strip at the bend and tie a peacock herl feather, then a small hackle feather on top of the foam. Wind thread forward followed by the ethafoam, creating a cone shaped underbody. On top of the foam layer, wrap the herl feather to just back of the eye and tie off. Tie in a small hackle feather just behind the eye and wrap twice, tie off and trim excess. Whip finish and glue.

MARABOU DAMSELFLY NYMPH

JANUARY/FEBRUARY 2000

During the month of June mature Damselfly Nymphs leave the security of aquatic vegetation venturing to surface and sculling their way towards shore to emerge. Mass emergences are common providing trout with a favourite food source perfectly silhouetted from above. Trout pursue the emerging Nymphs from the depths well into the shallows. Overcome by this Damselfly bounty marauding trout throw caution to the wind feeding with reckless abandon. Damselflies swim with an exaggerated Snake-like swagger, legs outstretched. The migrating Nymphs tire easy and take frequent rests. During an emergence position the pontoon boat or float tube so that casts are presented perpendicular to the shore. Retrieve the pattern back towards the emergent weeds. The mature Nymphs clamber out of the water to complete their transformation. Floating and intermediate lines work best, as the pattern must remain in the top foot of the water column. Use a steady hand twist or two- to four-inch strip retrieve featuring frequent pauses suggesting the many stops the Nymph makes along its emergence journey. Pursuing trout prefer to pounce as the Nymph pauses to rest.

MARABOU DAMSELFLY NYMPH

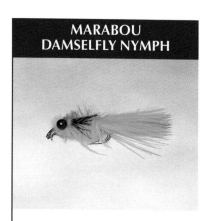

HOOK: #12 Tiemco TMC 2457 or Eagle claw LC56 Curved pupae or shrimp hook (juvenile imitation), #10 or #12 Tiemco TMC 5262 or similar 2X long nymph hooks (mature imitation)

TAIL/ABDOMEN/THORAX: Various colours of strung Marabou feathers (not Woolly Bugger Marabou) to match specific Nymph

RIBBING: Fine copper wire

WEIGHT: Fine lead wire

LEGS: Hungarian partridge rump feathers

SHELLBACK: Olive green synthetic raffia

EYES: Pre-made plastic or burned 20-pound test monofilament

MARABOU DAMSELFLY NYMPH

JANUARY/FEBRUARY 2000

INSTRUCTIONS: Form a thread base and tie in six to eight Marabou feather tips at the bend of the hook, extending at least as long as the hook shank, to form the tail. Leave the butts untrimmed as they will form the body. Tie in copper wire. Twist Marabou butts together and wrap the twisted mass two-thirds of the way up the hooks shank and tie down. Wrap the wire rib on in the opposite direction the Marabou was wrapped, tie off. Tie in the raffia wing case so it extends about half the distance down the hook shank. Add a few wraps of fine lead wire ahead of where the Marabou was tied off. Tie in several more Marabou feathers by the tips and build a thorax over the lead wire. Tie in four or five Hungarian partridge rump feathers on either side as legs. Bring raffia over thorax and legs and tie down. Tie in the monofilament eyes using figure-eight wraps. Bring raffia over the eyes and tie off fly.

BEAD-BODY CAREY SPECIALS

APRIL 2001

The Carey Special first appeared during the mid-1920s and was designed and popularized by Colonel Thomas Carey. It is one of British Columbia's best-known patterns and is still popular today as its charms still seduce trout. The pulsing pheasant rump hackle provides suggestive animation to the Carey Special. Just the trigger trout are looking for. In larger sizes the Carey Special represents Dragon Nymphs and Leeches. Some fly fishers feel a Carey Special is the best Dragon impostor out there. In sizes eight and smaller the Carey Special suggests immature Dragons and Sedge Pupa. The body colours are endless too, and many anglers keep a stash of red Careys as their, "shake em up" attractor fly when all other designs fail. The strip retrieve in partnership with a full sinking line or slow sink tip magnifies the pulsing characteristics of the Carey. A four- to six-inch strip retrieve consisting of slow to fast pumps and pulsates the hackle inducing takes from all but the fussiest of trout.

BEAD-BODY CAREY SPECIALS

HOOK: #8 Mustad 79580

THREAD: Matched to bead colour, or attractor colour

BODY: Two-millimetre outer diameter glass beads, colour varies

HACKLE: Bluish-grey, brown black or white pheasant rump hackle

INSTRUCTIONS: Pinch barb on hook and slide about eight glass beads on to the hook. If a bead will not fit over the hook, put it aside for a smaller pattern and select another bead. The glass will break if forced. Tie a small ball of thread over the hook at the bend to act as a bumper, match thread colour to beads, or choose a bright contrasting colour as an attractor. Make three half-hitches ahead of the wraps, and push beads back over them (providing room at the head for the hackle). Add a drop of Krazy Glue to secure the thread. Choose a large hackle feather and give two or three wraps, then wrap thread in front of hackle to push it back along the body. Whip finish and glue. Don't use pliers to grip the body of the fly when releasing fish, and beware when back-casting over rocky shorelines.

Nymphs

FLASH CHIRONOMID

HOOK: #10 to #16 TMC 2457
TAIL/BUTT: Red holographic Flashabou
RIBBING: Red wire
BODY: Pheasant tail fibres
THORAX: Ostrich herl
WINGCASE: Pheasant tail fibres
GILLS: Cream Midge Krystal Flash

EMERALD AND COPPER

HOOK: #8 to #16 Mustad C49S
THREAD: Olive Gudebrod 8/0
HEAD: Copper bead
RIBBING: Fine copper wire
BODY: Green holographic mylar
THORAX: Peacock herl
GILLS: White sparkle yarn

THE GUARANTEED

HOOK: #10 Mustad 9672
THREAD: UTC #70
TAIL: Teal
BUTT: Orange acetate floss, dipped in acetone for 15 seconds to give a hard shiny finish
BODY: Peacock herl
HACKLE/WING: Blue peacock neck feather

SELF CAREY SPECIAL

HOOK: #12 2XL
BODY: Brown fibres from a ringneck pheasant tail feather, wound to form a thin body
RIBBING: Fine gold tinsel, four turns
HACKLE: Immature ringneck pheasant saddle hackle, brown, two turns tied spider style, hackle should not extend beyond bodymore than body's length, preferably less

CRYSTAL SEDGE PUPA

MAY 2001

AQUATIC INSECT HATCHES RESEMBLE ICE-bergs, as the adults milling about are just the tip of the fly-fishing opportunities. Knowledgeable fly fishers recognize that the best parts of the hatch are hidden beneath the surface. Sedges or Caddis emergences are one such example. Weeks before the hatch Caddis Larva seal themselves within their cases transforming into the Pupa. Metamorphosis complete the Pupa cuts itself free shedding its larval case for the last time. Contrary to some, the Pupa does not skyrocket to the surface to emerge. As with other insects Sedge Pupa stage along the bottom gathering air and gases beneath the pupal skin to aid their pupal ascent and final transformation into winged adult. This staging process takes a few days and these lumbering Pupa do not go unnoticed by trout. Early in the hatch present the pupal pattern on or near the bottom with a slow hand twist retrieve, almost Chironomid paced. Wind drifting a pupal pattern is another ideal tactic. Watch the bow in the fly line for any sign of movement as many times trout pickup the pattern with little or no clue, often before the angler feels it.

CRYSTAL SEDGE PUPA

HOOK: #6 to #8 Mustad C49S
THREAD: Olive Gudebrod 8/0
BODY: Dubbed light-green Crystal Chenille and dark-olive Arizona synthetic peacock
WINGCASE: Brown raffia (Swiss Straw)
THORAX: Same as body
WING PADS: Mottled turkey quill
HACKLE: Partridge
ANTENNAE: Copper Krystal Flash
HEAD: Arizona synthetic peacock

CRYSTAL SEDGE PUPA

MAY 2001

INSTRUCTIONS: Begin by tying in a suitable length of Crystal Chenille. Next, form a dubbing loop about an inch shorter than the Crystal Chenille. Insert a dubbing hook into the loop and load the dubbing loop evenly. Pull the Crystal Chenille down parallel to the loop under the dubbing hook and back across the dubbing noodle. Begin spinning until the dunning fibres radiate perpendicular to the loop. Now, use hackle pliers to grab the bottom of the loop to keep it from unravelling. Hold the dubbing loop out directly behind the hook. Use a bodkin to fold the dubbing loop back across itself. Remove the bodkin and tie down the loop at the rear of the hook. The dubbing loop furls or twists around itself, forming an extended body over half to three-quarters of the shank. Wind the dubbing loop forward to the midpoint on the shank to form the rest of the body. Tie off the dubbing loop, but do not remove the excess. Fold it back out of the way and tie in the raffia wingcase. Unravel the raffia, cutting it vertically in two. Refold a section of the raffia so it is about one-eighth of an inch wide. With the wingcase in place, use the remaining dubbing noodle to create the thorax. Leave the two eye-widths of the shank clear to form the head. For the wing pads, trim a pair of slips from the midsection of a mottled turkey quill; a width slightly narrower than the hook gape is fine. Using a flexible cement, such as Dave's Flexament or Wapsi's Flex Seal, coat the slips and allow them to dry. Typically, this is the first step prior to the tying process. For mass-production, place the drying slips on wax paper to avoid bonding them to the tying bench. Tie the prepared wing slips along each side of the thorax so they angle down and back. Trim the slips just past the thorax to create the distinct wing pads of the natural pupa. With the wing pads in place, tie in a partridge hackle to suggest the pupa's gangly legs. Two wraps are fine. Avoid overdressing. Sweep the hackle down and back and tie it in place. Pull the wingcase over the thorax and tie it off. Tie in two strands of copper Krystal Flash at the head to create the antennae. Trim them even with the end of the body. Dab a small round head to complete the pattern.

Now that's a lunker!

THE WATER FLOATMAN

SEPTEMBER/OCTOBER 2001

As the annual Water Boatman fall progresses, trout that have seen a Boatman pattern or two can become discriminating critics of both pattern and presentation. Driven by instinct, trout focus upon a food source when it is most vulnerable. In the case of Water Boatman fall, they are most vulnerable at the surface. When Water Boatmen initially slam into the surface they scurry about as though knocked senseless from the experience. This spinning motion cuts through the surface film allowing the Boatman to escape into the depths. Fly fishers wishing to take advantage of the trout's focused behaviour should use a floating pattern that can be twitched and stripped at the surface to suggest the crazed sculling of the natural. Traditional patterns cast into the ring of the rise go ignored. In the absence of a buoyant Boatman pattern a Tom Thumb is a reasonable alternative. Floating Boatman patterns also work well in other applications such as working around aquatic vegetation. Less prone to snags and fouling buoyant boatman patterns allow fly fishers to probe areas that are not practical using other presentation techniques.

THE WATER FLOATMAN

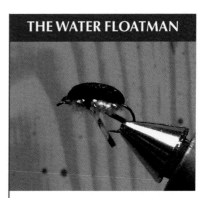

HOOK: #10 to #14 Mustad R 30

THREAD: Black Gudebrod 8/0

TAG: Silver or Holographic Silver Flashabou

SHELLBACK: Black Rainy's float foam or strip of black craft foam

BODY: Silver, Pearlescent, olive, brown or peacock Crystal Chenille

LEGS: Brown or olive Super Floss

THE WATER FLOATMAN

SEPTEMBER/OCTOBER 2001

INSTRUCTIONS: Beginning with a light-wire dry fly hook, such as the Mustad R 30, cover the front portion of the hook with tying thread. Tie in a length of silver or silver holographic Flashabou at the mid-point and, using close touching turns, wind the Flashabou down the hook into the bend. Wind the Flashabou back up the hook to the mid-point and tie off forming a neat, even tag in the process. The silver tag serves as a rallying point, suggesting the trapped air bubble these air-breathing insects carry with them on their sub-surface jaunts. With the tag in place, move the tying thread rearwards as it hangs halfway between the hook point and the rear of the crushed barb. Take a length of black Rainy's Float Foam and cut it in half in length to create two half-round pieces. Secure one of these prepared foam slips at the rear of the hook, flat side up for the shellback. With the shellback in place, bind in a length of Crystal Chenille at the rear of the hook and then advance the tying thread forward to the two-thirds point on the shank. Tie in a length of Super Floss for the distinctive oar-like legs perpendicular to the hook shank. Don't worry about the length, as they will be trimmed to size later on. Just make sure there is enough to trim. Wind the Crystal Chenille forward to create the body. Weave the chenille in and around the Super Floss legs so they stand out from the body and forward to the hook eye. Tie off the chenille at least one eye width back from the hook eye to avoid crowding the head. Trim away the excess chenille and then pull the foam strip over the back of the fly creating the shellback. Secure the shellback in place but do not trim the excess. Avoid pulling the foam too tight to keep away from diminishing any flotation. Lift the remaining foam strip that is protruding over the hook eye, build a neat head and whip finish. Trim the remaining foam strip even with the hook eye like an Elk Hair Caddis. Gather the Super Floss legs together and trim them even with the hook bend. be careful not to pull on the Super Floss tightly or a short, stubby pair of legs will result. If there are any doubts, settle for longer, more active legs.

EARLY SEASON BOMBER

MAY 2002

MUD BOTTOMED LAKES PROVIDE PRIME Chironomid real estate supplying the perfect environment in both numbers and size. Tunkwa, Leighton, Edith, are just a few of the lakes housing this Chironomid building block. Not surprisingly these lakes and others like them harbour dense populations of the larger species of Chironomid known by many as "Bombers." The pinnacle of size, the Pupa and Larva of these species reach sizes of close to one inch, supplying a rich source of protein for trout and char. Trout take Bomber Pupa with confident grabs. The Early Season Bomber is an ideal candidate to wind drift. From an anchored position quarter a cast across the surface. Allow the ambient wind to drift the pattern through the water column as it falls into the depths. The wind should put a gentle bow in the floating fly line as it sweeps around. Once the Early Season Bomber reaches the magic depth trout are feeding expect a firm take to this large morsel. Wind drifting is an ideal tactic, allowing anglers to cover water with a near static presentation. This method works for Chironomid Larvae, Caddis Pupa, and Leeches, practically any potential trout food source.

EARLY SEASON BOMBER

HOOK: #12 to #14 Partridge Klinkhamar Special GRS 15St
THREAD: Black Gudebrod 8/0
RIBBING: Small black v-rib and fine red wire
BODY: Silver Flashabou
WINGCASE: Pearlescent mylar
THORAX: Peacock herl
WING PADS: Burnt orange Super Floss
LEGS: Pheasant tail Angel Hair
GILLS: White sparkle yarn

EARLY SEASON BOMBER

MAY 2002

INSTRUCTIONS: To tie the Early Season Bomber, place the hook into the jaws of a vise and cover the shank with tying thread well into the bend. Return the thread to the 3/4's point on the hook and tie in a length of fine red wire. The redwire suggests residual hemoglobin many species retain from their larval youth. Next, take a piece of small black V-Rib and trim one end to a narrow spear point. Tie the prepared V-Rib in by the tip about halfway down the shank so the flat side of the V-Rib faces upward. This practice ensures the rounded side of the V-Rib faces outward on the completed pattern. Once the initial tie-in is locked in place, stretch the V-Rib to reduce bulk and secure in place along the hook shank. Return the tying thread up to the 3/4's point and tie in the silver Flashabou or a thin strip cut from an electrostatic bag electronic components are kept in. Secure the body material all the way down the shank to maintain a consistent slim-body profile. Wind the body material forward creating a slender even body, tie off at the 3/4's point and trim the excess. Wind the V-Rib forward in open spirals so the rounded side of the V-Rib faces out. Take the fine wire rib and follow the same path as the V-Rib laying the wire along the backside of the V-Rib. With the body and ribbing complete, tie in a length of Pearlescent Mylar for the wingcase. Advance the thread forward to the hook eye and tie in a sparse clump of pheasant tail Angel Hair on the underside of the shank so the tips protrude out in front of the eye. One the topside of the shank — tie in a strand of white sparkle yarn for the gills. Secure a piece of burnt orange Super Floss along each side of the thorax so the ends trail back along the side of the body. Tie in two strands of peacock herl and wind them forward to the hook eye, being careful not to crowd the head. Bring the Super Floss wing pads along each side of the Thorax and tie them off. Be careful not to pull them too tight or they can spring back out from under the securing wraps once trimmed. Sweep the pheasant tail Angle Hair back beneath the hook in beard fashion forming the legs. Trim the legs to length just past the thorax. Pull the wingcase forward, tie off and remove the excess. Trim the gills to proportion, about half the length of the thorax is fine. Whip finish and apply head cement.

Emergers

Most insects go through various transformations: from egg, to Larva, to Pupa or Nymph, and finally into an adult. Emergers are insects completing their metamorphous from Larva or Nymph stage to an adult form. This stage occurs at or very near the water's surface where the insect struggles to climb out of its nymphal exoskeleton. The exoskeleton is called a shuck once the insect has left it. Because of the emerging insect's vulnerability, trout will selectively search for and target this stage of an insect's development. The use of Emerger imitations can be very important to the fly fisher. Not all aquatic insects have an Emerger stage while in the water. Stonefly Nymphs, Dragonfly Nymphs and Damselfly Nymphs crawl out of the water before transforming into adults. However, having an adequate imitation of an emerging Mayfly, Caddis or Chironomid can be the difference between success and failure.

As an insect emerges, it will be half in and half out of its exoskeleton. Its wings will not be totally extended and the remaining exoskeleton will become translucent. The insect is usually struggling at this time, but not moving across the water. To imitate this stage, anglers should give their pattern a subtle twitch rather than a strip retrieve. The Tom Thumb is a classic example of an emerger. It can represent either an emerging Mayfly or an emerging Caddis.

Some Caddis and Mayflies emerge from their nymph stage below the surface and swim the remaining distance as a semi-transformed adult. This stage is also considered an Emerger. The Western Green Drake Mayfly is a good example of this. The wings are not fully extended and still need drying once on the surface, but the insect has already left its nymphal shuck behind. Small wet flies and the classic sparse partridge patterns are examples of Emergers that have hatched below the surface.

Most Emerger patterns have tails that represent the trailing shuck, and have wings that are shorter than adult imitations. Some patterns are designed to remain half in and half out of the water.

THE HALF AND HALF

APRIL 1992

SURVIVING THE RIGOURS OF THEIR EMERgence, ascent Chironomid Pupa hang at the surface in a comma like posture. Within seconds the Pupa alters position laying parallel to the surface. A split forms along the thoracic hump and the adult pulls itself free from the Pupal shuck, dries its wings and flies off. In most instances this process takes place in the blink of an eye. However if the emergence process is delayed trout abandon caution, honing in on the suspended Pupa. Fly fishers should have chironomid emergers that duplicate both emergence profiles, curved and horizontal. During intense emerger feeding trout may approve of one style forsaking all other offerings. This selective behaviour, frustrating to anglers, provides trout with an efficient method of feeding. Trout focused upon the curved comma profile scour the surface film choosing only those profiles and colours that match. Instinct has taught that all other profiles and colours do not qualify as food and as such are refused. At times trout switch preferences adding a further challenge for fly fishers to overcome.

THE HALF AND HALF

HOOK: #12 to #16 standard

TAIL: Soft fluff from the base of a dyed feather

BODY: Fine copper wire

HEAD: Fine deer hair

INSTRUCTIONS: Wrap the thread to the bend and fasten a short tuft of marabou fluff. Tie in and wind a section of fine copper wire in tight even wraps two-thirds of the way up the hook. Stack a small amount of deer hair and tie it in with the butts towards the eye of the hook. Fold the deer hair back over itself and tie off tightly at the eye. The hair should flare out around the eye. Whip finish behind the flared head and add a drop of glue.

GREEN DRAKE EMERGER

JUNE 1996

Many fly fishers believe that all aquatic insects emerge upon the water's surface. In reality, this is not the case as the final transformation from aquatic form to winged adult manifests itself in a number of ways. Insects such as Dragonflies, Stoneflies and Damselflies complete their final transformations by crawling out of the water. A few species of Mayflies and Caddis use this same method. The famed Green Drake Mayfly uses a unique emergence process, transforming beneath the surface and breaking through the surface film as the sub imago or dun. During surface emergences trout dining upon adults can become ruthless critics of both presentation of pattern and presentation. Trout shunning all dry fly approaches are often deceived with a damp presentation using an emerger pattern. Casting the pattern upstream to establish the correct drift, lift the rod and apply line tension raising the pattern into the path of the feeding trout. In many instances this approach out performs the more widespread dry fly approach. Should trout be rising throughout the length of a run raise, lower and feed in fly line as possible to repeat this process through a single drift.

GREEN DRAKE EMERGER

HOOK: #10

TAIL: Wood duck flank feather (or dyed mallard flank)

RIBBING: Copper wire

BODY: Dubbed grass-green Polydub or similar synthetic

WING: Grey Z-Lon or dyed Phentex wool

LEGS: Wood duck flank

HEAD: Bronze peacock herl

INSTRUCTIONS: Wrap thread to bend and tie in a short clump of flank feather for a tail. Attach wire rib above tail. Form a dubbing loop and paint the thread with a sticky wax, then dub on a neatly tapered clump of body material. Wind on a fat body three-quarters of the way up the hook. Cover with four wraps of copper rib and add a short vertical wing. Fasten five or six strands of flank feather to either side to create the legs, then make another turn-and-a-half of dubbing and tie it off. Add the head of peacock herl, then whip finish and glue behind the herl.

THE COPPER EMERGER

MARCH 1997

After surviving the perils of ascent emerging insects are not out of the woods. Depending upon conditions the final push through the surface tension may prove to be greater obstacle than running the gauntlet of the ascent. On slow moving or stillwaters flat calm conditions provide a barrier many insects are unable to conquer. Parallels have been drawn comparing the emergence process to digging through six feet of dirt. On damp or choppy days many insects are swamped at the critical point of emergence. No matter the reasons, numerous insects become entangled within their shucks. Unable to escape, stillborns stuck and quivering are easy pickings for surface feeding trout. Placing specific stillborn patterns in front of fussy risers is often the only approach for consistent success. Stillborn patterns incorporate a variety of tail materials to suggest Nymphal and Pupal shucks. Hackle and partially spent wings simulate toppled adults unable to escape the clutches of the surface tension. When trout are at the surface sipping stillborns and emergers watch the rise forms. Concentrated pods of fish working a small area are often the smaller age class. Keep a keen eye out for a lone delicate rise. Lone rises often belong to the largest trout. Educated by experience these elusive feeders are tough to spot and challenging to catch.

THE COPPER EMERGER

HOOK: Short shank #18 to #10

THREAD: Dark olive

TAIL: Four strands of grizzly hackle, three strands of Krystal Flash

ABDOMEN: Fine copper wire

THORAX: Bronze peacock herl

HACKLE: Dry-fly-quality grizzly hackle

INSTRUCTIONS: Tie in the tail, equivalent in length to the body of the fly. Start copper wire slightly below the bend of the hook and wrap to midway point of shank, then tie off the wire. Fasten two strands of bronze peacock herl, twist around thread and wrap in a short thorax, leaving a space for the hackle. Tie in grizzly hackle and make at least six tight wraps to the eye of the hook. Tie off, trim, whip finish and glue.

M&M

JUNE/JULY/AUGUST 2001

WET FLIES ARE MOST OFTEN THOUGHT OF AS Emerger patterns, but wet flies span other situations as well. All aquatic insect adults return to the water to lay eggs, fostering another generation. This activity signifies the end of their lives. Exhausted and spent they drift downstream. Churned by the current, waterlogged adults are swept beneath the surface providing an easy meal for the trout without the risks of surface feeding. Wet flies and emergers provide fly fishers the means to take advantage of this opportunity. Wet flies and emerger patterns work during subsurface egg laying activity. Females of certain aquatic insects such as some species of Caddis and Mayflies dive or crawl beneath the surface, attaching their egg clusters upon submerged vegetation and debris. First time observers to this activity might be fooled, believing the rolling and splashing trout might be chasing hatching adults. Study the rise forms carefully and the specific behaviour of the insects before making a snap judgment. If necessary, walk into the water and seine the water for diving adults using an aquarium net. Keep an eye out on the surface for shucks as further confirmation of a hatch.

M&M

HOOK: #14 or #16 Mustad 94840 or Signature series R50, Eagle Claw L 59 or Tiemco 5210

THREAD: Moss green Danville 6/0, 8/0 or equivalent

TAIL/BODY: Six to ten strands of mallard breast feather

OVERBODY: Invisible mending thread

WING: Rolled mallard breast fibres

HACKLE: Grizzly hackle

INSTRUCTIONS: Wind the thread to the rear of the hook and tie on a 20-centimetre length of invisible mending thread. Tie in the tail using long strands of mallard breast feather extending one-third the length of the hook beyond the bend, do not trim the butts as they will form the body. Wind the thread to the head and follow by winding the butts of the tail strands to the head and tie off. Wind the overbody to the head in the opposite direction the body was wound. Use 10 to 12 strands of mallard breast and roll them to form a wing, tied in one-quarter the length of the hook back from the eye, extending to the bend. Tie in hackle and make one or two turns, keeping it short and sparse. Whip finish and glue.

Knots

76 Knots

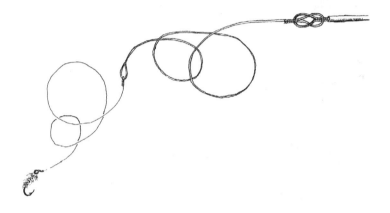

Knots

You can have the right line, the right fly, the right rod and reel. But, the knot you tie can be your weakest link. Putting a knot in a line puts a weak spot in the line as the knot is now the weakest section. For the maximum strength possible, reliable knots are a must. The ideal knot will retain strength in the line as if there were no knot tied.

There are only few knots that fly fishermen really need to know. Those are: the Clinch or Improved Clinch knot, the Blood knot, the Surgeon's knot, the Tube, Nail or Needle knot and a Loop knot.

The Clinch or Improved Clinch knot is used most frequently when tying fluorocarbon or monofilament line to a hook. The Improved Clinch is far less likely to slip when tying with monofilament. The single Clinch seems to break less frequently when using fluorocarbon..

The Blood knot and Surgeon's knot are used when tying two lengths of leader material together. The Surgeon's knot is easier to tie and reported to be the stronger of the two, but only when properly tied. The Blood knot is neater and leaves both sections in a straight line.

The Tube, Nail or Needle knots are all similar and used primarily when attaching monofilament to fly lines. The advantage of the knot is it buries the tip end under itself. It is also useful when attaching two non similar lines together.

A Loop knot is needed when a fly needs to swing freely. There are several good ones including the Duncan Loop knot, and the Double Loop.

Remember to use the right knot for the right application. With nylon monofilament, knot failure is usually due to improper tying while with braided and microfilament lines, implementing the wrong knot can cause knot failure. So go ahead and tie one on and happy fishing!

BLOOD KNOT

1. Overlap the ends of the two lines. Take one end and twist it four times around the other line. Then bring it back and pass it between the two lines.

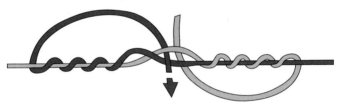

2. Repeat with the other free end, taking care that the first does not unravel.

3. Wet the knot to lubricate it, then pull it tight and trim off the ends.

IMPROVED CLINCH KNOT

1 Pass the line through the eye of the hook.

2 Make five turns around the line.

3 Thread the end through the loop ahead of the eye.

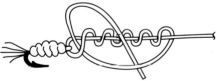

4 Bring it back through the large hoop.

5 Pull firmly to tighten.

DUNCAN LOOP

1. Thread line through the eye.

2. Fold back loose end, then fold back again to form a second loop.

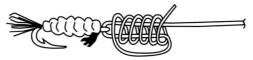

3. Wrap loose end five times around both parts of the first loop, always passing inside the second loop (away from the eye).

4. Pull to tighten.

NAIL KNOT

1 Lay nail butt end of leader, and tip end of fly line parallel, with fly line headed left and leader right.

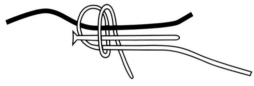

2 Wrap leader snugly five or six times.

3 Pass the butt end of leader through middle from left to right.

4 Pull tight and remove nail. Tighten and clip ends.

SURGEON'S KNOT

1 Double the end of line and make an overhand knot.

2 Leave loop open and pass end of line through a second time.

3 Pull the line to tighten and trim ends.

Dry
Flies

84 Dry Flies

Dry Flies

Dry flies are basically patterns that sit on the surface and imitate the adult insect. Keeping the fly pattern on the surface requires the use of buoyant materials to overcome the weight of the hook. Most dry flies use lighter wire hooks. Buoyant materials could be from one or more sources. There could be closed cell foam on the body, or the use of deer or elk hair that has a hollow core. Stiff, bushy hackle will keep some flies on the surface and so will fur dubbing that holds tiny air pockets. For the fly tier, the difficulty is using enough materials to keep the fly floating while at the same time still imitate the slender bodies of many insects.

Closed cell foam has the advantage of being light, tough and very buoyant, but it is difficult to shape and colour properly. It is ideally suited for bodies on terrestrial patterns, but it is too opaque for Mayfly patterns. It is okay for bodies on the larger Stoneflies and Caddis.

Deer and elk hair is hollow and makes an ideal wing material for Caddis patterns or Stoneflies. Deer hair will flare to almost 90 degrees when bound tightly in one spot. When it is cut, spun, stacked and trimmed it makes a fat, buoyant body. Tightly stacked deer hair can be trimmed to just about any shape.

The stiff, shiny fibres of a rooster's neck hackles resist water and support a dry fly. The rougher the water, the more fibres are needed to support the fly. Too many fibres change the appearance of the fly. The tier needs to know what the fly is imitating and where it's likely to be used. As a general rule, bushy patterns are used on rough water and sparse patterns are used in calm water.

Natural fur, or synthetic dubbing retain tiny air pockets that help in the buoyancy of a fly. When treated with a silicone floatant the dubbing resists water and helps keep a fly on the surface.

ROYAL WULFF

FEBRUARY 1981

Christened the strawberry shortcake of fly patterns, the Royal Wulff is another of the must have dry fly patterns. The Royal Wulff is perhaps the most famous pattern within the late Lee Wulff's stable of flies. Lee Wulff took the venerable Royal Coachman and beefed it up for chasing Atlantic salmon and Brook trout in Eastern Canada. The swapping of white quill wings and pheasant tippet tail for calf tail and elk hair respectively provided the magic ingredients for a buoyant durable combination. The Royal Wulff's generic look — its peacock and floss body draws trout to the fly when all others fail. During the heat of the day when nothing appears or is hatching is a prime time for an attractor style pattern such as the Royal Wulff. Many anglers use the Royal Wulff as a searching pattern in the absence of anything obvious. Probe those areas that offer trout comfort, protection and availability to food. Preying on the trout's opportunistic and aggressive disposition, the Royal Wulff is a proven trout magnet.

ROYAL WULFF

HOOK: #10 to #16 Mustad 94840
WING: White calf tail
TAIL: Moose body hairs
BUTT/SHOULDER: Peacock herl
BODY: Fine red floss
HACKLE: Two dark brown, dry-fly quality hackles

INSTRUCTIONS: Tie a small clump of calf tail on top of the hook to form a wing the same length as the hook shank. Taper the butts with scissors. Divide the hairs into two equal portions on either side of the hook, creating a "V". Cross the thread four times through the two wings and under the shank. Then make two wraps around the base of one wing, and two around the base of the other, finishing with one more cross wrap. Add thinned head cement to the base of the wing to secure it. Wind the thread in tight wraps to the bend and tie in 12 to 20 moose hairs to form a tail as long as the hook. Wind on two peacock herls for four turns. Then two strands of floss to just past the halfway point, then two more herls as before. Remember to wind the thread forward between each segment so each can be tied off, then wind the thread to the eye. Tie in two hackles with the face away from you. Wind one to the eye, making half the wraps behind and half in front of the wing. Follow with the second winding through the first in the same fashion. Form a small head with half hitches and add a drop of head cement.

TOM THUMB

APRIL 1981

The Tom Thumb is synonymous with B.C. fly-fishing.

This versatile pattern has arguably duped more surface feeding trout than any other. By varying the quantity and colour of deer hair the Tom Thumb represents just about any airborne insect from caddis to ants. For years a green-bodied size 12 Tom Thumb worked wonders on Skagit River trout. Debates have raged that the Tom Thumb looks nothing like the insects it does such a wonderful job imitating. Closer observation might prove otherwise. As a sedge emerges the Pupal shuck trails behind while the insect crawls forward to shed the last vestiges of its aquatic form. As soon as the wings are free they are pushed forward and upward to dry and stretch them out. At this instant the Tom Thumb is a dead ringer for an emerging sedge. A vulnerable stage trout instinctively hone in on. The fan wing of the Tom Thumb is a superb visual cue for young and old eyes alike. Simply cast the fly and wait for a response often works best. When this heave-and-leave tactic wanes try giving the fly the odd strip or twitch to ring the dinner bell.

TOM THUMB

HOOK: #10 or #12 Mustad 94840 or equivalent

THREAD: Green monocord

TAIL: Mule deer hair

SHELLBACK/WING: Mule deer hair

INSTRUCTIONS: Tie in a tail of well-marked, medium dark mule deer hair at the bend to form a tail about the length of the hook shank. Bind firmly to the hook shank with a generous number of wraps and clip off the excess. Select a large clump of deer hair, about three times longer than the hook, align the tips and secure it on top of the hook shank, butts extending forward, tips trailing back. Wind the tying thread tightly to the hook eye, back to the tie in point and back again to the eye. Trim the excess at the eye from the top of the hook only. Pull the clump of trailing hair over the back of the fly, leaving the tail undisturbed. Hold the material tightly forward and make several tight wraps just in front of the underbody, then take two turns and release the material as you tighten, allowing the tips to flare. Lock the newly formed wing in place with a few turns forward of the wing to hold it vertical and above the lateral line. Finish with several half hitches.

ELLIOTT'S IRON BLUE HUMPY

AUGUST 1984

There are a number of core patterns within any dry fly box. The Adams, Blue Dun, and the Royal Wulff are a trio that comes to mind. With the multitude of freestone rivers and streams present in the west the Humpy or, as it is know in some circles, the Goofus Bug is one pattern all western dry fly buffs should carry.

Owing to the buoyant materials used to construct Humpies they are an ideal selection for the rough and tumble runs of riffles and pocket water. The Humpy's unique combination of moose, deer hair and hackle provides an appeal trout can't resist. Some Humpies have even been elevated to royal status, mimicking the colour scheme of a Royal Wulff.

The trick to proportional Humpies is tying in the correct length of deer or elk hair. Some find elk hair easier to work with while making for a more visible Humpy. Stack the hair to even the tips and then tie it in at the mid point of the shank. The deer hair should be a length equal to the tip of the tail to the eye of the hook.

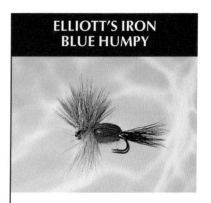

ELLIOTT'S IRON BLUE HUMPY

HOOK: #10 to #18 Mustad 94840 or equivalent

BODY: Dubbed blue-grey muskrat underfur

SHELLBACK/WING: Deer hair

HACKLE: Two dry-fly quality hackle feathers, one grizzly, the other Iron Blue Dun

---

INSTRUCTIONS: Tie in about a dozen stacked moose mane hairs for the tail, ensuring that their length is the same as the total hook length and that they are in line with the hook. Tie a stacked clump of deer hair of the desired bulk and colour in at the bend, the tips trailing back exactly twice the tail's length. Dub a body of muskrat hair about three-quarters of the way up the hook. Pull the deer hair over the back to the thread and tie it down firmly. Raise the remaining deer hair tips to vertical and lock in place with thread wraps in front of the wing root. Done properly, the wing height will match the hook length. Tie both hackles in by the butts, dull sides facing you, behind the wing. Wind one half of the first hackle length behind the wing and the rest in front. If the first hackle had four turns behind and three in front, reverse that for the second hackle, three behind and four in front, to balance the fly. Finish with several half hitches.

PALE MORNING DUN

THE WALHACHIN GREEN

HOOK: Standard Dry fly size 14-20
THREAD: Pale yellow
BODY: Pale yellow dubbing
TAIL: Light dun hackle barbs
HACKLE: Light dun
WING: Medium dun hen hackle tips

HOOK: #10 or #12 thin wire
BODY: Green poly dubbing
WING: Dark, fine deer hair

INSTRUCTIONS: Attach the thread and wind it to the hook bend. Dub a sparse, slim body of green poly yarn to the eye of the hook. Fasten a short clump of dark, fine deer hair on top of the hook to form the wing, extending back about the length of the body. Trim the wing butts and form a small head, whip finish and coat lightly with head cement.

THE SALMON CANDY

JULY 1984

The Salmon Candy is an Oregon based series of patterns all designed to represent various stages of a caddisfly's transition from Pupa to adult.

Of the four stages two and three suggest the emergence phase of this cycle. Trout like most predators are opportunistic feeders preying upon the weak and helpless. During the emergence process all aquatic insects are at the mercy of the elements. Committed to transition unable to flee. Despite the presence of adult sedges seasoned fly fishers know the value of emerger patterns and their ability to coax wary surface feeders. To the inexperienced the myriad of rise forms and fluttering adults tricks the fly fisher into a brisk decision to fish dries. Slow down. Watch the rise forms of foraging trout to detect emerger feeding. Emerger feeders tend to show only their backs as they vacuum their hapless prey from the surface film. Time the rise and place the greased Salmon Candy into the trout's path. Allow the Salmon Candy to sit motionless bobbing within the surface film. The rise is often deliberate and subtle. A simple raise of the rod sets the hook. Even if trout are dining on the odd adult emerger patterns take the majority of trout.

PRE-EMERGENT SALMON CANDY SEDGE

HOOK: #8 or #10 Mustad 9671 or 97831
TAIL: Deer hair
BODY: Olive green wool
SHELLBACK: Deer hair
HACKLE: Palmered brown hackle, trimmed top and bottom

INSTRUCTIONS: Tie in a short, sparse tail of deer hair at the hook bend. Tie in a small, stacked clump of deer hair by the tips, also at the bend. Then add a brown hackle feather, also tied by the tip, followed by a length of body wool. Wind the wool to the eye and tie off, follow by palmering the hackle to the eye as well. Clip the hackle fibres top and bottom, then pull the shellback deer hair that was tied in at the bend over the back and tie in at the eye, leaving a small exposed clump of deer hair for the head. Whip finish and glue.

THE SALMON CANDY

JULY 1984

EMERGENT SALMON CANDY SEDGE

HOOK: #8 or #10 Mustad 9671 or 97831
BODY: Olive green wool
HACKLE: Brown
WING: Deer hair
THORAX REMNANT: Deer hair

INSTRUCTIONS: Attach a length of wool at the bend and wind it about two-thirds the way up the shank, tie it off but do not trim. Select a small, dry-fly quality hackle feather and tie it in at this point. Wind the hackle like a collar for three or four wraps, then pull the hackle fibres down, under the body and lock them in place with two turns of the thread. Align the tips of a small segment of deer hair and tie it in as a wing, leaving enough hair to form the thorax remnant. Continue wrapping the wool to the eye and tie off. Pull the deer hair left over from the wing over the new wool wraps and tie down securely at the head, leaving a short stubby head of deer hair butts. Whip finish and add a drop of glue to the wraps.

THE CALLIBAETIS MAYFLY

APRIL 1985

There are many species of Mayfly that call stillwaters home. Here in B.C. Callibaetis is the species of consequence.

On the interior lakes Callibaetis provide the dry fly aficionado their first real opportunity of the season. Emergence begins from late May on low lying waters and continues through spring into summer. The first hatches of the season on a given body of water are the largest, typically in the size 12 range. As the season progresses the prodigy of the first hatches has a reduced growth cycle and anglers will notice a steady decrease in dun and spinner size as the season progresses. By the end of the Callibaetis hatch cycle size 14 or even 16 duns and spinners may be the norm. Observant fly fishers aware of this trend carry a suitable complement of patterns to carry them through the season. Favoured Callibaetis dressings, such as the Callibaetis Mayfly, should be demure in profile and lie flush upon the surface. Callibaetis frequent clear marl lakes, environments where trout can be fussy and difficult to coax to the fly. Leaders must be light and on calm conditions degreased to avoid marring the surface. Fluorocarbon, so popular with subsurface techniques is a disadvantage as in many instances it drags the fly beneath the water.

THE CALLIBAETIS MAYFLY DUN

HOOK: #14 or #16 Mustad 94840 or 94833 fine wire

TAIL: Three or four moose leg hairs

BODY: a mixture of poly dubbing, three-quarters tan, one-quarter olive

RIBBING: Dark grey moose mane

WING: Dyed grey-blue deer hair

HACKLE: Grizzly hackle

INSTRUCTIONS: Tie in a tail of moose leg hair, curving out to the sides to help keep the fly upright. Tie in the moose mane rib hairs, then dub on a body of poly dubbing mixed three-quarters tan, one-quarter olive or as naturals dictate. Tie off body and add a wing of deer hair, tied so that it flares away from the hook and tilts slightly back towards the tail. Tie in a hackle feather, make two or three wraps and tie off. Whip finish and glue, then trim the hackle fibres from below the lateral line.

THE CALLIBAETIS MAYFLY

APRIL 1985

THE CALLIBAETIS MAYFLY SPINNER

HOOK: #14 or #16 Mustad 94840 or 94833 fine wire

TAIL: Three or four moose leg hairs

BODY: A mixture of poly dubbing, three-quarters tan, one-quarter olive

RIBBING: Dark grey moose mane

HACKLE: Grizzly hackle

WING: White deer hair or clear plastic poly-film wrap (plastic bags), folded and cut out as a pair

INSTRUCTIONS: Tie in a tail of moose leg hair, curving out to the sides to help keep the fly upright. Tie in the moose mane rib hairs, then dub on a body of poly dubbing mixed three-quarters tan, one-quarter olive or as naturals dictate. Tie off the body and add a wing of deer hair, clipped top and bottom, with body dubbing wrapped figure-eight style between the hair. Alternatively, a pair of wings can be made out of folded clear plastic, cut out with scissors or a wing-cutter. Puncture the plastic many times with a needle to give the illusion of the original. After the wing is tied in, tie in a hackle feather. Make two or three wraps of the hackle, tie off, whip finish and glue.

THE PARACHUTE HACKLED FLY (MOSQUITO)

MAY 1985

PARACHUTE PATTERNS OFFER MANY ADVANtages to the fly fisher and have replaced traditional hackled patterns for many dry fly anglers. Selective trout demand the best of the fly angler in both presentation and pattern. The parachute pattern provides a slim realistic profile and the circular spread of the hackle suggests the surface footprint of the natural insect. Parachute patterns also have a practical side as they always land right side up. Not nearly as proportionally demanding as traditional dries, fly tiers can afford a degree of latitude at the vise. Any food source can be imitated with a parachute design from Grasshopper to tiny Chironomids. Many traditional patterns also benefit from a parachute transformation. Patterns such as the Adams, Royal Wulff and the Mosquito benefit from a parachute make over. The Mosquito is a generic pattern that does a good of job not only imitating its namesake but other insects including Chironomids and Mayflies. At home on both still and moving waters the Mosquito's versatility serves the fly fisher well.

THE PARACHUTE HACKLED FLY (MOSQUITO)

HOOK: #12 to #18 Mustad 94840

THREAD: Black

TAIL: Moose mane hair

BODY: Moose mane hair, one light, one dark

WING: Teal or mallard flank feather

HACKLE: Grizzly

INSTRUCTIONS: Tie in a tail of four or five moose mane hairs, equaling the body in length. Then tie in two moose mane hairs, one light and one dark, by their tips. Wrap forward covering two-thirds of the hook shank and tie off. Cut equal portions from each side of a heavily-barred mallard flank feather to form the wings, lay them back-to-back so they flow away from each other, and tie them in. Allow a lot of excess wing material to remain. Pull the wing forward and make several wraps behind it to hold it upright. Tie in a hackle of the appropriate length, by the butt, underside in. Grasp it with your hackle pliers, and wind it around the base of the wing. Hold the wing firmly with your fingers and use a hackle guard to keep the fibres in place. Tie off and lacquer the base of the wing heavily, then shape the head and whip finish. If the wing has been forced back, grab the wing butts you left long and work it back into position before the lacquer sets, then trim the excess.

THE SIMPLE SEDGE

JULY 1985

Despite the collection of complex intricate fly patterns available today it is refreshing to know simple is still best. During a concentrated Sedge emergence "dry fly fever" is a common consequence. Trout swirling and smashing Sedges all around the boat creates target overload for the uninitiated. Too many choices cause frustration as casting strokes match the heartbeat while feet trample line. Compounding the fever, light is failing and a sense of, "It will be over before it starts!" grips the fly fisher. The last thing needed to is a fly pattern that doesn't float or isn't durable. The Simple Sedge deer hair construction provides supreme buoyancy and durability. With the fly pattern in the anglers favour emotions can be harnessed. Choose one fish and cover it with a well-placed cast into the ring of the rise. Wait for a response. If there are no takers strip the fly to induce a take. Repeat this process and focus upon the task at hand to avoid excitement from overtaking. As the light fades reduce leader length. Using a disciplined stroke make a cast and strip the Simple Sedge back to the boat. Under the veil of darkness the eyes take a back seat to ears as the distinct splash of the take is immediately followed by the firm pull of a rainbow.

THE SIMPLE SEDGE

HOOK: #8 2X long dry fly
TAIL: Deer hair
BODY: Deer hair
WING: Deer hair

INSTRUCTIONS: Tie in a small clump of deer hair at the tail. Gradually add more clumps of hair, spinning them into place as you move up the shank towards the head. Tie off when you can add no more hair to the shank. Clip the hair closely on the bottom of the body leaving the back one big wing akin to a Matuka-style fly. Trim the head closely all the way around the shank.

Lucky fishing hat.

STIMULATOR

JULY 1987

The Stimulator is one of the best rough water patterns available to the fly fisher. Created by renowned fly fisher and tier Randall Kaufman, the Stimulator has wide application through out B.C. On those rivers with healthy populations of large Stoneflies such as Salmon flies (*Pteronarcys*) and Golden Stones (*Acroneuria*), the Stimulator excels. However, don't limit this pattern's exposure to Stonefly waters. By varying the size and colour anglers can use Stimulators to suggest Caddis, including the famous Traveler's Sedge, and Hoppers. The deer hair wing does an excellent job of representing the fluttering wings of a Stonefly adult as it hops and skips across the water's surface. This pattern can be skittered or presented dead drift. As the drag free portion of the presentation ends don't immediately pick up and recast. Allow the pattern to swing at the end of the drift pulling it under. Trout follow the pattern as it sweeps around, don't be surprised by a strong grab as the fly hangs below. Another tactic involves fishing the pattern damp in the surface film. At times trout smash such offerings with reckless abandon.

STIMULATOR

HOOK: #6 (or smaller) Mustad 9672
THREAD: Orange-brown
TAIL: Deer hair
BODY: Yellow (rear two-thirds) and orange (front third) polypropylene yarn
HACKLE: Palmered brown (rear two-thirds) and grizzly (front third)
WING: Deer hair

INSTRUCTIONS: Attach the thread, wind it to a point over the barb and tie in a stacked, fairly substantial clump of deer hair for the tail. Strip the left side of a brown saddle hackle and tie it in by it's tip at the tail, tie in a length of yellow yarn at the same place. Wind the yarn to the two-thirds point and follow by palmering the hackle in even wraps, tie off and clip both hackle and yarn. Stack and tie in another clump of deer hair that extends almost to the tail for the wing. Trim the wing butts to a forward tapering ramp, saturate in head cement and tie down securely. Tie in a dry fly quality grizzly hackle, by its butt, good side away from you, and a length of orange yarn on top. Wind the yarn to the eye and tie off. Follow with the hackle as a full dry fly hackle but with a moderately open spiral wrap. Tie off, whip finish and glue.

THE CHOPAKA MAY

JULY 1987

Chopaka Lake is located in eastern Washington and for hardened Callibaetis chasers it is a body of water held in high regard. British Columbia has many similar Callibaetis hot beds including Lac Le Jeune, Lac Des Roche and Big Bar. American angler Boyd Aigner's Chopaka May was created with such bodies of water in mind.

Callibaetis duns seem to be preferred trout food in B.C. Callibaetis display a marked preference for emerging on dreary overcast days. Begin by searching the chara and marl shoals Callibaetis prefer with demure nymph patterns. Around midday attention switches from sub surface fare as the speckled duns pop out dotting the surface. Within minutes anglers are surrounded by a Callibaetis regatta along with trout more than willing to drag a few below. With so many naturals to choose from trout set up predictable beats tipping up every few seconds to draw in a dun. Make a calm controlled cast ahead of the predicted rise, strip out any remaining slack and wait for the acceptance rise.

THE CHOPAKA MAY

HOOK: #12 to #20 Mustad 94840, 94842 or equivalent

TAIL: Moose body hair

BODY: Moose mane, one dark and one light

WING: Deer hair

HACKLE: Dark blue dun or dun/grizzly mix

INSTRUCTIONS: Start by tying in a wing of stacked deer hair, tips facing forward, about the length of the hook shank. Lift the wing to vertical and secure with several wraps of thread and a drop of head cement. Trim the butts of the deer hair so they taper down to nothing at the bend, then wrap the butts with thread. Tie in two long moose hairs for the tail and separate with a criss-cross of the thread. Then tie in on dark and one light moose mane hairs. Wrap thread to the wing and follow with the moose mane pair, creating a segmented light and dark abdomen. Tie in two dry-fly quality hackle feathers with their good sides away from you, sized so their fibres will extend about three-quarters of the wing height. Wind the first hackle four turns behind the wing, then four turns in front. Follow with the second wound carefully through the first. Tie off, whip finish and glue. Cut a "V" section from the hackle under the fly, to help it sit upright on the water.

WESTERN MARCH BROWN

APRIL 1988

On rivers and streams throughout western North America the March Brown symbolizes the kick off of a new fly-fishing season. In southern latitudes the first significant emergence of the season starts in March as its namesake suggests. Here in British Columbia our hatches begin about a month later. The Cowichan River on Vancouver Island provides anglers with one of the best opportunities to encounter March Browns. Preferring to emerge in the swifter flows of a river, anglers must have buoyant patterns and sound presentation techniques. Drag must be eliminated to ensure an accurate presentation. The reach cast is an ideal technique for controlling drag. Once the forward cast of the delivery stroke is made simply reach the rod up stream perpendicular to the current flow. The fly line "reaches" upstream from the fly to the rod tip presenting the pattern at the current's pace ahead of the fly line. Hold a few feet of fly line in reserve and pop the rod tip vertically to feed the slack into the drift prolonging the presentation. Don't be too aggressive with the popping action as the fly should not move during.

WESTERN MARCH BROWN

HOOK: #12 or #14 Mustad 94833 or equivalent

WING: Natural or slightly dyed blue-grey deer hair

BODY: Tan poly dubbing

HACKLE: Cree or mixed ginger and grizzly hackle

TAIL: Four strands of moose leg, or beaver or otter body hair

---

INSTRUCTIONS: This fly starts with the deer hair wings. Hold a small clump vertically on either side of the hook and wind the thread tightly around the base, then fold some of the butt portion back under the hook and wrap it with thread very tightly. Trim the excess and add a drop of glue. Tie in a tail using four moose leg hairs, slightly fanned out from one another. Use a dubbing loop to wind on a thin body of poly-dub to the wind, then let it hang and tie in one or two Cree hackles. Wrap one turn of dubbing around the base of the wing, one turn in front of it and tie off. Wind the hackles like a thorax in front and back of the wing and tie off. Whip finish and glue. Trim the hackle under the hook into a "V" to help the fly sit upright.

DEER HAIR CADDIS

JUNE 1988

IN THE WORLD OF CADDIS OR SEDGES HERE IN the commonwealth the Traveler's Sedge raises the eyebrows of any angler familiar with British Columbia's stillwaters. Images of large insects scampering across the surface in a vain attempt to flee disaster come to mind. Large confident rises and swirls mark the final resting place of these rich morsels.

As their size would indicate the Traveler's Sedge is a robust insect that rides in the water rather than on it. Successful fly patterns need to run almost decks awash to be successful. The low riding profile of the Deer Hair Caddis provides an ideal silhouette for trout to hone in on. Hook ups are positive as trout inhale the pattern with ease.

When fishing large dry flies keep the leader between nine and twelve feet, the clearer the water the longer the leader. Tippet should be stout to permit proper turnover without alerting suspicion. Should one have to resort to lightweight tippets, pinch the fly line as it shoots towards the target in mid air. This tactic slingshots the fly so it does not fall back upon itself, eliminating slack that seasoned trout are quick to exploit.

DEER HAIR CADDIS

HOOK: #6 or #8 2X long

BUTT: Pale-green antron

ABDOMEN: Mixture of grey and green antron or seal fur

WING: Mixture of dyed brown and natural grey deer hair

HACKLE: Cree or a combination of grizzly and brown

HEAD: Spun and clipped natural deer hair

INSTRUCTIONS: Wind the tying thread to the bend and dub on a small portion of pale-green antron. Use a dubbing loop with mixed proportions of grey and green fur, wind it up the body three-quarters of the way up the hook, tie off. Mix some dyed brown and natural deer hair in a stacker and align the tips, and tie it on top of the last bit of dubbing so that it does not flare too much. Tie in hackle feather(s), then spin a deer hair head and clip it close to the hook. Wrap the hackle over the head and tie off. Whip finish and glue.

FOAM CALLIBAETIS EMERGER

HOOK: #10 to #16 Tiemco 2487
THREAD: Tan 8/0 Uni Thread
TAIL: Mottled turkey flats or partridge
RIBBING: Pearlescent Crystal Hair
BODY: Tan super-fine dubbing
WING: Grey poly yarn
WING CASE: One-quarter centimetre grey foam strip
LEGS: Grizzly saddle, palmered through thorax
THORAX: Tan super-fine dubbing

INSTRUCTIONS: Tie in four strands of tail material, two forking to each side of the hook shank. Tie in one strand of pearlescent Crystal Hair, then form a dubbing loop and wind the dubbed body material from the bend until even with the point of the hook. Follow with the ribbing and tie off, continue dubbing forward to the hook eye. Tie in the wing, extending to the bend. Tie in the foam wing case and a small hackle feather. Hold the wing case, palmer the hackle in a few evenly spaced wraps to the eye and tie off. Trim top and bottom. Fold wing case forward and tie down behind the eye. Whip finish and glue.

PARA PUPA

HOOK: #10 to #16 2487
RIBBING: Pearlescent Crystal Hair
BODY: LaFontaine Touch dubbing mix
WING: Fluorescent Dan Bailey's Hi Vis or poly yarn
THORAX: Peacock herl
HACKLE: Grizzly hackle (wound parachute style)

INSTRUCTIONS: Wind the tying thread to the tail and form a dubbing loop. Use the dubbing mix and wind on a body, stop one-third the hook shank from the eye. Tie in a peacok herl, spin in the thread and wind on the thorax stopping just short of the eye. Next tie in a wing of Fluorescent Hi Vis or poly yarn, extending just past the bend of the hook. Tie in a hackle feather and wind on parachute style around the wing for two wraps, tie off and trim excess. Whip finish and glue.

THE GREEN DRAKE

JULY 1988

REVERED THROUGHOUT THE FLY FISHING community, the famed Green Drake hatch draws anglers to hallowed waters such as the Henry's Fork of the Snake River in Idaho. Here in British Columbia, fly fishers also have the opportunity to sample this hatch on their home waters. The Skagit River is one such opportunity. Green Drakes have a unique emergence process, as the transformation from Nymph to dun takes place sub surface. This wrinkle in the Green Drake's emergence process makes emergers the pattern of choice. Allowing a dun imitation to sink and swing at the end of the drift allows for the best of both worlds — dry and wet. If a trout has taken up a specific feeding station, anglers can apply tension to sink the pattern at the trout's nose. Green Drake's are a large insect by Mayfly standards; trout seldom refuse a dead drifted dun. The large size 10 and 12 duns flutter and bounce upon the surface as they attempt to dry their wings prior to permanent flight — a commotion few trout can resist. Although the emergence window for this hatch is limited, trout keep an eye out for this rich food source.

THE GREEN DRAKE

HOOK: #10 or #12 dry fly hook
TAIL: Dun fibres from a moose leg
RIBBING: Yellow or brown floss
LEGS: 50/50 mixture of dyed grey and green deer or elk hair
WING: Dyed slate grey body feather of a partridge or duck

INSTRUCTIONS: Begin by shaping a number of wings with a wing burner. Tie a matched pair of wings in an upright position and then fasten securely to the hook. Tie about 10 strands each of dyed grey and green deer hair in under the hook, tips aligned and facing forward, at the base of the wing. Wind the tying thread over the hair butts to the bend and then back to the eye, clip excess. Spread the deer hair legs evenly to either side of the hook and hold them in place with figure eight wraps of tying thread and glue the wraps. Tie in four or five moose leg hairs for the tail. The curve of the hair should spread out to either side of the hook. Tie in a strand of floss, and form a dubbing loop of poly-dub. Wrap the thread to the eye, follow with the dubbing, then the floss wrapped in the opposite direction. Make on more wrap of dubbing, tie it off, whip finish and glue.

TERMITE

SEPTEMBER 1988

Termites are another example of an acute terrestrial food source that, for the most part, trout have little opportunity to see. But on those bodies of water adorned with adjacent timber and rotting wood, termites are a welcome change from the normal fare. In the late summer, migrating Termites can fill the air, creating feeding frenzies on nearby waters. Termites are poor, clumsy fliers whose flight path often takes them over water with no hope of crossing. Exhausted the Termites plummet to the water below, wings outstretched, providing a large silhouette for trout to hone in on. As with many food items, trout can become fixated on colour, and successful Termite patterns should mimic the brown-orange colouration of the naturals. On lakes nestled amongst known Termite populations, look for the leeward points containing fallen rotten timber. Termites venturing from the nest are swept downwind and trout in the area take up station. Anchor the boat, float tube or pontoon boat down from the point. Make a perpendicular and slightly downwind cast and wait for the trout to respond. Twitch the pattern to suggest a struggling termite. Don't overdo the twitch, too much puts trout off the pattern.

TERMITE

HOOK: #10 Mustad 94833 or equivalent fine wire
BODY: Mixed 2:1 orange and brown poly-dubbing or similar synthetic
WING: Natural dun spade or saddle hackle
HEAD: Same as body

INSTRUCTIONS: Wind the tying thread to the bend and paint with sticky wax. Spread the 2:1 mixture of orange to brown dubbing material, form a dubbing loop and spin it with heavy hackle pliers. Wind the fat abdomen halfway up the hook make a thin body for about one-quarter of the hook and tie in a hackle feather. Wind on the hackle, ensuring it remains centred and tie off. Spread the barbules to the sides of the hook and hold them in place with figure-eight wraps. Wind on a smaller head of the same dubbing and tie off. Whip finish and glue. Clip any stray hackle fibres from the top and bottom. Grease the fly with a good flotant before use.

MAYFLY

MARCH 1989

The Mayfly must have been created for the dry fly enthusiast. With its unique two-stage adult phase consisting of dun (sub imago), and spinner (imago), Mayflies provide multiple opportunities to dupe trout. In the case of the March Brown, there is much literature regarding duns but little credited to spinners. Spinner falls draw trout from all points, attracted by the rhythmic dancing and dapping of the gravid females. The rigours of mating and egg laying are lethal to Mayflies, with most having adult life cycles measured in hours, not days. Ian Forbes, originator of the Mayfly, noticed that trout seemed to show more appeal for the spinners prior to their death than after. Traditional spinner designs feature spent wings synonymous with dead or near dead spinners. Ian's hackle wing suggests the still active spinner as it first touches the water. As with all Mayfly adult presentations, dead drift is the name of the game. Introducing controlled slack into the equation is the key to drag free floats. When positioned above a feeding trout, try the serpentine cast. On the final delivery stroke, wiggle the rod tip from side to side just prior to the fly line touching down, causing the line to land in a serpentine fashion, fly at the lead.

HACKLE-WING MAYFLY

HOOK: #14 Mustad 94840 Partridge

TAIL: Dun hairs from the lower leg of a moose

BODY: 50/50 mixture of orange and brown fine poly dubbing

HACKLE: Finest quality medium blue dun neck hackle

INSTRUCTIONS: Tie in four dun coloured moose leg-hairs at the bend to form the tail, ensure they are spread out to help support the hook. Spin a dubbing loop using a 50/50 mixture of orange and brown poly dubbing. Keep the dubbing sparse to maintain a slim, Mayfly body profile. Wrap the dubbing three-quarters of the way up the hook shank to represent the wing. Wind the hackle on tightly and tie off. Moisten fingers and pull all the hackle fibres above the hook, then pull the remainder of the dubbing under the hackle and make two turns at the head. No hackle fibres should remain below the hook shank when completed. Whip finish and glue.

THE DEER HAIR GOLDEN STONE

APRIL 1989

Golden Stones (Acroneuria) are probably the most widespread Stonefly species found in British Columbia. Just about every region in the province boasts a river or two with a healthy population. Rivers such as the Cowichan, Stelako, Thompson and St. Mary's all come to mind. Creatures of oxygenated riffle-water, Stonefly adults offer the best opportunity for the fly fisher as they return to the waters surface to lay eggs. At this time opportunistic trout gorge themselves silly on these large, clumsy fliers. As this activity progresses and trout have fallen to the basic Stonefly patterns, a degree of selectivity occurs. Trout become exacting critics of both pattern and presentation. Low riding patterns typically work best and the Deer Hair Stone supplies an appealing profile for the most critical of feeders. The Deer Hair Stone is not just a pattern for the trout angler — many of British Columbia's famed Steelhead rivers harbour healthy Golden Stone populations. More than one summer run Steelhead has succumbed to the charms of a properly presented Golden Stone imitation.

THE DEER HAIR GOLDEN STONE

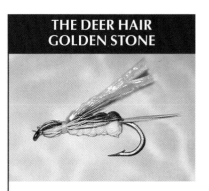

HOOK: #4 to #8 3X long
TAIL: Dyed goose biots
BODY: One-third green, one-third brown, and one third yellow deer hair or moose mane
RIBBING: Yellow floss
WING: Plastic raffia
HEAD/LEGS: Same as body

INSTRUCTIONS: Tie in a thick clump of white or pale grey deer hair at the tail, then a section of yellow floss. Wrap the thread two-thirds of the way up the hook and tie in a mixture of green, yellow and brown very long deer hair from the tail. Tie in the hair at the butts with the tips pointing towards the bend, then wrap the thread in even turns to the bend of the hook and back again. Fold the deer hair back on itself, spreading the hair around the hook, and tie off. Try to keep the yellow on the bottom the green in the middle and the brown on the top of the fly, then wind the floss over the body in even wraps. Make the head using shorter hair with the same technique, but tied in the opposite direction, tips facing the eye. Leave some of the tips of the hair flared out to the side to represent legs, selectively trim off the rest. Whip finish behind the head and coat the whole fly in varnish.

DANCING CADDIS

JUNE 1992

The late Gary LaFontaine's classic, Caddisflies is a must read for any fly fisher. Gary's insights towards trout behavior, Caddisfly habits and pattern design have provided a foundation for fly fishers worldwide. The Dancing Caddis is one example of Gary's unique insight towards pattern design. Utilizing the Partridge Swedish Dry Fly hook, the Dancing Caddis provides a unique alternative for fussy trout. As with the Interior Caddis, the Dancing Caddis rides inverted with the hook point masked within the construction of the fly. In addition to camouflaging the hook, this design allows fly fishers to skitter their pattern across the water to simulate the egg-laying dance of many Caddis species, particularly those common to rivers and streams. To skitter and dance the Dancing Caddis, lay a quartering cast across the current. As the pattern begins its downstream drift, raise the rod tip to dance and skitter the pattern. Lower the rod tip to introduce slack line, and then repeat the process. Skittering is a deadly tactic when egg-laying females or adults are returning to the water for an evening sip. Trout chase and crash the surface in an attempt for an evening snack — providing an exhilarating experience.

DANCING CADDIS

HOOK: #10 Partridge Swedish
BODY: Dubbed green fur
WING: Deer hair
HACKLE: Ginger brown

INSTRUCTIONS: Use a dubbing loop from the bend of the hook to wind on a body of green fur. The dubbing loop should cover the hook to the peak of the kink on the Swedish hook. Tie off and turn the hook upside down in the vise. Attach a pencil sized clump of stacked deer hair to the peak of the kink, forcing the hair down to each side and holding in place with wraps of thread. Next, attach a good-quality ginger hackle to the peak of the kink and wind it over the base of the deer hair. Tie off at the eye, whip finish and glue. Trim the hackle away from the bottom of the kink so that all the hackle fibres flare upwards.

MICHAEL CAMP'S CADDIS

NOVEMBER/DECEMBER 1992

Seasoned fly fishers have a number of strategies for coping with trout that have seen just about everything — or least their behaviour suggests they have. One successful strategy involves the selection of unorthodox or non standard patterns. Michael Camp's creative use of CDC and unique wing construction makes his Caddis pattern an ideal candidate for both selective and non-selective feeders. The realistic profile of this design makes it perfect for flat water conditions on both moving and stillwaters. The scruffy looking CDC feathers offer an appealing meal for the most discriminating trout. The naturally buoyant CDC fibres provide needed flotation and gather air bubbles to further sway trout in the angler's favour. When trout are rising anglers should position themselves to allow close, accurate casts without spooking their quarry. The preferred approach involves stalking the trout from below and to the side taking advantage of the trout's narrow blind spot. Keep a low profile, kneeling if necessary. Avoid false casting over the target. Cast away from the trout and then make the final presentation cast on target.

MICHAEL CAMP'S CADDIS

HOOK: #6 to #12 Partridge Roman Moser spear point

BODY: Olive, green or brown and orange fine poly dubbing

TAIL: Natural grey Cul de Canard (CDC) feather

HACKLE: Palmered natural grey Cul de Canard (CDC) feather

WING: Brown, spotted feather glued to fine dressmaker's netting

INSTRUCTIONS: Select speckled feathers of grouse partridge or turkey for the wing. Fasten them to a fine dressmaker's netting using a thinned, flexible glue. Allow them to dry then trim to shape and fold into the tent-like shape of a Caddis wing. Wind thread to the bend and fasten a CDC feather near its tip, allowing some of the feather to form the tail. Let the remainder of the feather hang, and form a dubbing loop. Apply sticky wax and dubbing material, then wind a thin body forward, then palmere the CDC feather over the body like a hackle using evenly spaced wraps. The dubbed body should be visible through the wraps of CDC. Attach the wing, then make two more wraps of CDC feather over the head. Whip finish and glue.

GRIFFITH'S GNAT

MARCH 1993

Considered by many to be an essential pattern, the Griffith's Gnat is the consummate tiny dry fly. Ideal in sizes 16 and smaller, the Griffith's Gnat seems to lose its appeal in larger sizes. Tied originally to simulate clustering midges on small spring creeks by American George Griffith, founder of Trout Unlimited, the Griffith's Gnat is perfect for suggesting any minute food source, including Ants and Beetles. On many streams it is common to observe trout rising to minuscule food sources almost too small for the human eye. The natural reaction might be to tie on something large to play the greed card. But for trout that have witnessed too many presentations this tactic falls short. Under these conditions, downsize the tippet and tie on a size 18 or smaller Griffith's Gnat. Trout rise with confidence sipping down the Griffith's Gnat without hesitation. Use a slack line presentation, such as the puddle or serpentine cast, to permit a long drag free float. The Griffith's Gnat must coast as though unencumbered by fly line or tippet.

GRIFFITH'S GNAT

HOOK: Captain Hamilton, 1X, No. 18-22

BODY: Peacock herl from just below the eye of the tail feather

HACKLE: Grizzly, stripped on one side and palmered

INSTRUCTIONS: Pick a well-marked grizzly hackle from near the top of the neck. Choose a hackle with barbules no longer than the bend of the hook. Strip the barbules from one side of the hackle, and flair out the remaining barbules by gently stroking them away from the shank. Fasten the tip of the hackle near the bend of the hook, then tie in a strand of peacock herl at the same place. Use extra-fine, dark-olive tying thread. Waxed thread is preferable because it stays where it's placed. Twist the peacock herl around the thread, then wind both together around the hook shank. Wrap the peacock herl and thread to just behind the eye of the hook and tie it off. Wind the grizzly hackle over the peacock body, in the opposite direction to the way you wound the thread. Make no more than four or five wraps of the hackle, or the fly will appear too bushy. Tie off the hackle, whip finish, and add a tiny drop of glue to complete this fly.

Fishing from canoe.

THE GRASSHOPPER

JULY/AUGUST 1993

During the heat of summer when most hatches have come and gone, terrestrials such as Grasshoppers take centre stage. Terrestrial insects have no aquatic stage at all, yet they inhabit the brush and grasses along the shores of every river and stream. With a jump first, look second philosophy, Grasshoppers are in constant peril along the stream bank. Trout inhabiting rivers and streams with long grassy banks soon become accustomed to this hapless fare crashing onto the water's surface. In many instances it is the crash landing of the Hopper that draws the trout's attention. Using a short, stout leader of nine feet, use a heavy-handed approach to presentation by taking advantage of this dinner bell response and slap the pattern onto the surface. Another approach involves deliberate over casting. If possible, drop the pattern onto the far bank and then strip the pattern back so it plops naturally upon the surface. If there are no takers after the initial presentation, drift twitch the pattern activating the rubber legs. The natural action of the rubber leg material does a superb job duplicating the feeble struggles of the soon to be water logged Hopper and is often the trigger trout seek.

THE GRASSHOPPER

HOOK: 2X long, No. 10 to 6

BODY: Ethafoam cut to shape and coloured with marker pens

WING: Microweb coloured with marker pens

HEAD: Deer hair, bullet shaped — top green, bottom yellow

LEGS: Rubber band behind head

THE GRASSHOPPER

JULY/AUGUST 1993

INSTRUCTIONS: Start by laying a base of a yellow tying thread along a shank of the hook to the bend, the cover it with a pliable glue. Slice the body of ethafoam part way through with a razor blade, and slide it down to cover the hook's shank. A short section of ethafoam should extend slightly beyond the bend of the hook, and the remainder should reach two-thirds up the hook shank. Wind the tying thread over the ethafoam in four even wraps. keep the thread firm, but not so tight that it will compress the foam too much. Cut a wing from micro-web wing material and mark it with a grey Pantone pen. Add a drop of pliable glue to the top of the ethafoam and place on the micro-web wing. Hold it in place with several wraps of tying thread. Make a bullet head of dyed deer hair. Use green for the top and yellow for the bottom. Fasten the deer hair with the tips forward and the butts along the hook shank. Leave the tying thread at the base of the foam, then fold the green deer hair back over the top of the hook and tie it off where it meets the foam. Repeat the process with the yellow deer hair under the hook. Use the cut-off tip of a ball point pen to shape the deer hair into a smooth head. Trim the tips of the yellow deer hair so they don't mask the ethafoam body. Leave the green deer hair tips long, so they will assist in landing the fly properly. the final stage is to tie in the rubber legs. Mark the rubber leg material with an orange pen and tie an overhand knot in each piece. Fasten the legs one at a time so they sit properly. One piece of rubber does both the front and back legs. Thread compression will force the legs out to the side so they will wiggle on the water. Whip finish where you tied in the legs, then add glue to complete the hopper.

GODDARD SEDGE

FOAM BEETLE

HOOK: #14 to #18 Tiemco 100 or similar fine-wire, dry fly hook
BODY: Spun deer hair
HACKLE: Dark brown dry-fly
INSTRUCTIONS: Spin on a body of deer hair until about three-quarters or the shank from the bend. Tie off deer hair body and trim to shape, leaving a short stubby tail of some longer bristles at the bend. Tie in a hackle feather and wind several times and tie off. Whip finish and glue. Trim the hackle from below the hook, leaving two strands angling away from each other to represent legs. Colour body dark-brown with a waterproof marking pen.

HOOK: #10 to #16 Tiemco 100 or 101
SHELLBACK: Foam strip
BODY: Peacock herl
INSTRUCTIONS: Tie in the foam shellback at the hook bend. Tie in a peacock herl just forward of the point where the shellback was attached. Spin herl in the tying thread and wind forward to just short of the hook eye. Pull shellback over and tie in at the eye. Whip finish and glue. Rubber legs can be threaded through the body to complete the fly if desired.

MIKULAK SEDGE

MAY 1995

British Columbia fly fishers owe a debt of gratitude to the late Art Mikulak and his perseverance in creating the consummate Sedge pattern. Art's triangular concoction of elk hair, seal's fur and hackle has created legions of loyal followers. At the height of a Sedge hatch, the Mikulak Sedge in is tough to topple. Trout take this pattern without hesitation, at times gulping it before the natural bobbing next door. When stripped, the Mikulak creates a convincing wake, capable of drawing trout from a considerable distance.

Position the cast so the fly is retrieved upwind. The textured surface quickens the emergence process and sedges seem to pop up everywhere. Trout not afforded the luxury of a prolonged emergence are forced to chase and smash the pattern. It is not unusual to have trout boil at the fly in attempt to swamp it and slow it down. Avoid the temptation to strike, cease the retrieve and wait for the trout's return. It often takes a few occurrences before realizing what is going on.

MIKULAK SEDGE

HOOK: #6 to #12 Tiemco TMC 5215 (2XL, fine wire) or Mustad 9671(2XL)

THREAD: Olive green

TAIL: 10-12 pieces of stacked elk hair, about one-third of shank length.

BODY: Tightly dubbed synthetic seal fur

WINGS: Light-coloured elk hair

HACKLE: Dark brown

INSTRUCTIONS: Tie in a short tail, form a dubbing loop and apply the dubbing, twisting it tightly. Wrap the dubbing halfway up the hook shank then add 15 to 20 pieces of stacked elk hair, extending back almost to the tips of the tail. Continue dubbing half the remaining distance to the hook eye, and add another clump of stacked elk hair, whose tips should extend almost as far as those of the first wing. Tie in hackle and continue dubbing to the eye, tie off. Wind hackle two turns and tie off, then trim the top and bottom hackle fibres. (Note: on smaller flies like the #12 2X one wing is enough for floatation purposes).

RED ANT

SEPTEMBER/OCTOBER 1995

ANTS ARE THE PEAK OF THE TERRESTRIAL PIE, and so are the imitations created to match them. On lakes, Ants are of limited importance, but as with most focused food sources, when Ants slide onto the menu trout can become extremely selective. Ants run the full gamut of sizes from the large three-quarter-inch Carpenter Ants to minuscule size 24 Cedar Ants. On lakes, Ant feeding can become focused. Trout gorge themselves to excess turning off the bite for days after as they struggle to digest their excess. Ants are rich in formic acid and are challenging for fish to digest. With no antacid in sight, digestion is slow and fly fishers suffer. Rivers and streams see a steady diet of Ants throughout the season, especially during the late summer when ant activity is at its highest. Resident trout keep a keen eye for ants sprinkling down from overhanging bushes, branches and trees. Ant patterns are an ideal hatch-buster on rivers. When trout appear immune to the best of Caddis, Stonefly or Mayfly imitations, a lowly Ant pattern drifted through pods of feeding fish often draws positive response. Use slack line presentations to create drag-free drifts. Keep a watchful eye for the gentle sipping rise common to an Ant pattern.

RED ANT

HOOK: #14 or #16
THREAD: Black monocord
BODY: EVA foam
HACKLE: Brown

INSTRUCTIONS: Prepare foam bodies, by trimming cylinders slightly with sharp, fine scissors or a razor blade so that the ends are slightly larger than the middle. Slice the body lengthwise about two-thirds up the length of the cylinder. Lay a base of thread on the hook shank, paint with flexible glue, and press the foam body on so that the abdomen hangs off the back. While the glue is setting, fasten the body with thread and tie off. Colour the bodies using felt pens (mostly red abdomen, mostly black head) and then apply at least two coats of lacquer to create a nice shine. Retie the thread to the middle of the hook and tie in a brown hackle, giving it two turns before tying off. Whip finish and glue. Trim the hackle from the bottom of the hook.

YELLOW SALLY

SEPTEMBER/OCTOBER 1995

EVERY RIVER AND STREAM FLY FISHER'S FLY BOX should feature a Yellow Sally pattern of some description. This species is widespread in British Columbia, providing many opportunities for anglers to put impostors to work. The smaller *Alloperla*, while not as well known as their larger cousins *Acroneuria* and *Pteronarcys*, are not ignored by trout. Needing a constant flow of oxygenated water to survive, anglers should focus their stone fly efforts in the riffle areas of streams. Riffles provide ideal camouflage masking the fly fishers presence, allowing close accurate casts. The agitated water requires repeated presentation, as trout do not always see the offering on the first pass. Break the riffle into small manageable chunks and fish each segment in a methodical fashion. Drift the pattern in amongst the stones and debris, paying special attention to those areas that offer respite to trout. Expect bold, aggressive rises as trout have little time for complex decisions. If the presentation is sound, response should be quick and positive.

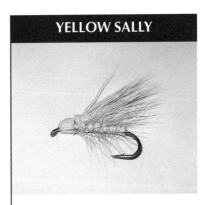

YELLOW SALLY

HOOK: #12 Mustad 94840 or 1X long #14 Mustad 9671
THREAD: Yellow
BODY: Pale yellow deer hair
HACKLE: Bright yellow hackle
WING: Same as body

INSTRUCTIONS: Fasten a small clump of deer hair by the butts at the centre of the hook. Cover the hair to the bend with tight wraps so it will not spin, and paint the covered section with head cement. Fold the remaining hair forward, enveloping the entire hook shank in deer hair. Hold the hair while giving five even wraps it to keep it in place, forming a body two-thirds up the hook shank. Tie off the deer hair tightly and add a drop of glue. Tie in hackle by its butt and continue winding thread forward over the remaining hair. Tie off the hair tightly at the eye, but don't trim it. Wind thread back to where the hackle was tied in and tie in a short wing of deer hair, so that the stacked tips extend just beyond the end of the hook. Trim excess butts. Fold excess hair at eye back over hook towards the wing, enveloping it again. Tie off and trim excess. Wind hackle behind the bullet shaped head, tie it off, whip finish and glue over the deer hair body.

CICADA

APRIL 1996

RELEGATED TO CAMEO APPEARANCES WITHIN the terrestrial cast, Cicadas still manage an arrival or two upon the water. Spending up to 17 years in a sub-terrainean larval form, Cicada emergences are not legendary. Nonetheless, when one of these large squat-appetizers plops into the water, the trout's opportunistic tendencies take over. Not a widespread phenomenon, there are regions of British Columbia, such as the Kootenay's, were the Cicada patterns are a worthy addition to the dry fly box. The Columbia River is one location within B.C. all anglers should book a trip to. The huge back eddies of this powerful river create Lazy Susan feeding setups where trout can go on Cicada binges. Pounding the banks with Cicada dressings is another sound tactic. Angler positioning is key in order to take full advantage of the swirling currents. Back eddies can be plied by casting blind or scouring the converging currents for the subtle signs of rising trout. If possible, stalk trout from behind — placing the cast to the side and above. Allow the pattern to drift back down towards the fish.

CICADA

HOOK: #10 Mustad 9671 or equivalent
BODY: Pre-made closed cell foam
WING: Stacked white deer hair
LEGS: Olive dyed moose mane
THORAX: Dubbed dark green Flashabou

INSTRUCTIONS: Lay a good base of thread on the hook shank and coat with glue. Slice the underside of the pre-cut foam body slightly and press it over the glued hook. Attach the foam in two places with strong tying thread, bound tightly to prevent the body from spinning on the hook. Colour the foam black and cover it liberally with head cement, after drying apply a second coat. Re-attach thread at the rear connection where the foam was attached, and tie in a spare wing of stacked deer hair. Add three strands of moose mane to each side of the body. Form a dubbing loop, sticky wax it and dub in a small amount of green Flashabou. Wrap thread over the top of the foam and around the moose hair at the second junction. Wind three wraps of Flashabou dubbing over the middle section of foam and the front stub of the wing. Whip finish and glue. Trim the dubbing closely and the legs evenly to complete the pattern.

THE BEETLE

MAY 1998

THE BEETLE

BEETLES FORM THE LARGEST MEMBERSHIP OF all insect orders, and despite their land borne ways, Beetles are popular with fly fishers and trout. Although found in gargantuan sizes, it is the squat, oval size 12 and smaller Beetles that slide onto the trout's menu most often. As with Grasshoppers and Ants, Beetles end up in and on the water during the course of their natural activities. Feeble and clumsy once, wet Beetles float in the waters surface, wriggling like mad in a vain attempt to escape. As with most terrestrial imitations, Beetle patterns work best during the heat of summer. Choose suspected trout lies near or under overhanging branches or bushes. Shady areas are tough to prospect but are just the places where beetles fall from into the jaws of the trout below. Probe these areas by casting the fly upstream above the suspected lie or overhanging bushes. Pinch the fly line, swinging the pattern back towards the angler. When the Beetle pattern crosses in line with the bushes, let the fly go and feed in slack. Done correctly, the pattern drifts with precision under the branches without the need of the fly fisher having to cast it there. This method works for placing patterns in around and under other obstructions.

HOOK: #10 to #16 Tiemco 100 or 101
SHELLBACK: Foam strip
BODY: Peacock herl

INSTRUCTIONS: Tie in the foam shellback at the hook bend. Tie in a peacock herl just forward of the point where the shellback was attached. Spin herl in the tying thread and wind forward to just short of the hook eye. Pull shellback over and tie in at the eye. Whip finish and glue. Rubber legs can be threaded through the body to complete the fly if desired.

FISHER'S GNAT

JANUARY/FEBRUARY 2002

Jim Fisher's Gnat is an excellent example of suggestive fly tying, providing fly fishers with a pattern capable of different missions on still or moving waters. An evolution of the traditional hackled dry fly, the parachute styling of Fisher's Gnat furnishes a realistic silhouette capable of fooling the most discriminating feeders. Less proportionally specific, parachute patterns always land in a graceful upright fashion. On stillwaters trout can inspect a fly with ruthless scrutiny. The natural footprint of Fisher's Gnat is often all the proof wise trout need to feed. Under ideal conditions, frauds and naturals disappear in unison as Rainbows take full advantage of the emerging duns. On the Skagit River, located a few hours east of Vancouver, trout have become conditioned disciplined feeders, demanding an exacting presentation. A tiny size 16 or smaller parachute pattern, such as Fisher's Gnat, drifted drag-free on a 6X leader over visibly feeding trout is the order of the day. Larger, more traditional offerings tend to be met with a slow ascent and follow before the trout retreats to the depths in obvious disapproval.

FISHER'S GNAT

HOOK: #12 to #16 Mustad 94840 or equivalent

THREAD: Black nylon 6/0 or 8/0

TAIL: About 10 web free fibres from a grizzly saddle or neck feather

BODY: Two to three long peacock herl and fine copper wire

WING: White or fluorescent hot orange, red, or chartreuse calf tail

HACKLE: Dry-fly quality grizzly hackle

INSTRUCTIONS: Lay and even thread base from hook eye to bend. Tie in tail materials and trim butts. Tie in body material and wrap to mid-shank, tie off. Hold remaining body material out of the way with a material clip. Tie in a small, sparse clump of stacked wing material by the butts, then trim butts and cover with thread. Build a thread base on the eye side to stand the wing up. Circle the wing base to build a firm post. Tie in hackle by its butt, then wind the remaining body material to just back of the eye, tie off and trim excess. Attach wing to a gallows tool (if you use one) and wind the hackle four to six turns around the wing post. Tie off hackle tip, trim excess, whip finish and apply clear head cement.

LADY McCONNELL

MAY 2002

CREATED BY FAMED B.C. FLY FISHER, BRIAN Chan, the Lady McConnell is the ideal choice when trout are fixated on adult Chironomids. Chironomid adults return to the water, laying their eggs during the low conditions of morning or evening. At this time, the water is calm and the risk of avian predation is reduced. Most fly fishers witness to these egg-laying flights find trout infuriating with their selective gentle rises. There are two strategies: Leave the Lady McConnell free to float, or begin a slow hand twist retrieve if the trout are focused upon the adults scurrying along the surface. Match the "V" wakes of the naturals to draw a response. Takes can be subtle, almost imperceptible. Be observant, target trout using predictable rise patterns and cast the fly where the next anticipated rise should be. This strategy is one of the most exciting forms of fishing, as the trout may or may not rise again. As light fails, raise the rod at any rise in the vicinity of the fly. It is amazing how many times takes of the fly are not seen or felt.

LADY McCONNELL

HOOK: #12 to #16 Mustad 94833

THREAD: Black 6/0 or 8/0

TAIL: Grizzly hackle tip over white fibres

BODY: Dubbed muskrat fur or appropriate coloured tying thread

OVERBODY: Deer body hair

HACKLE: Grizzly

INSTRUCTIONS: Cover rear two thirds of hook shank with thread. Tie in tail material keeping tail sparse and grizzly hackle shiny side up, trim to shank length. Tie in deer hair tips two thirds up the hook shank and secure rearwards with the tying thread. Dub a slender body. Pull over deer hair to form shell back. Tie in hackle feathers. Wind hackle forward. Build head and whip finish and apply head cement.

Salmonids

122 Salmonids

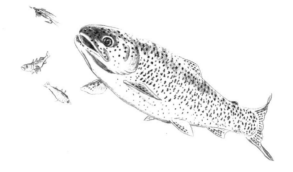

Salmonids

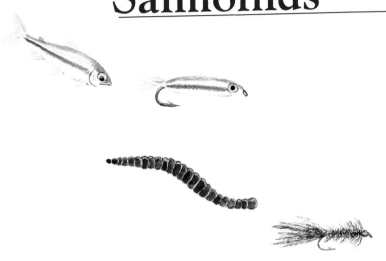

Streamers and bucktails are the patterns most often used when fishing for cutthroat trout, steelhead and salmon. Streamers and bucktails usually represent baitfish or leeches. Streamers have feather wings and bucktails have hair wings. One problem with streamer patterns is the long feathers have a tendency to twist under the hook and cause the fly to spin. The stiffer hair wings on bucktail patterns do not do this as often.

Marabou or the soft aftershaft plumes found at the base of body feathers add fish catching appeal to many streamer-type flies. Many baitfish imitations and leech patterns consist of nothing but the soft marabou plumes. The soft marabou plumes in the Woolly Bugger's tail move easily in the water and suggest life. Variations of the Woolly Bugger can represent many different things and makes that pattern one of the most effective flies anywhere.

Tube flies are flies tied on hollow tubes of plastic or metal. They are used when very large or extra long patterns are needed, but without the disadvantages of oversized hooks. They have the advantage of allowing for a change of hook if it gets damaged. It also means the hook sits near the tail end of the fly where the fish most often strikes. Tying flies on tubes requires specialized equipment to hold the fly in the vise.

WOOLLY WORM (BLACK)

MAY 1980

Fly fishers on the hunt for Steelhead spend a good portion of their time searching for prime Steelhead water. As a general rule Steelhead prefer holding in water three to six feet deep with a walking pace current. Variables to consider in this basic equation include bottom structure and makeup. Obstructions that break up the current provide ideal hiding spots for resting Steelhead. Large basketball sized rocks shelter Steelhead from the constant pounding of the current. Bottom depressions are favourite haunts that are difficult to detect. Steelhead, as with most salmonids, prefer not to hold in swift flows. Narrows and other reaches within a river that force Steelhead to work through the current often have slower relaxed stretches above and below the confluence. Resting Steelhead take full advantage of these resting areas. Fly anglers identifying these areas find sweet rewards by methodically working their patterns through all promising areas. Slow near-stagnant waters are not favoured Steelhead lies, as the oxygen content within these waters does not provide adequate Steelhead comfort.

HOOK: Mustad 9672 #10-#6 long shank
TAIL: Fluorescent Steelhead yarn
HACKLE: Palmered brown saddle hackle
BODY: Black chenille

INSTRUCTIONS: Wind thread around hook and leave hanging even with hook point. Quarter a short section of yarn lengthwise and tie in a one quarter section back of thread, trim forward section as close to thread as possible, trim rear facing yarn even with outside bend of hook to form tail. Take a long thin grizzly hackle and run fingers from tip to butt to make fibres stand out from quill. Tie in the tip of the feather with a few wraps even with hook point, then wrap to eye, leaving thread just back of the eye. Pull 1.5 centimetres of fuzz from a 15 centimetre section of chenille and tie in de-fuzzed core where tail and hackle were tied in. Wind chenille to a point just back of the hook eye (0.3 centimetres), tie off and clip excess to complete the body. Palmer hackle in open spirals, tie off just back of the hook eye and finish with a drop of head cement.

THE DAVIE STREET HOOKER

FEBRUARY 1983

STANDING ON THE BANKS OF A B.C. STEELhead stream, the first order of business might be line choice, floating or sinking. Art Lindgren, one of British Columbia's best known Steelhead fly-fishers, uses a formula based upon water depth and temperature. Runs and pools over six feet in depth dictate sunk line presentations. This depth can decrease in the event of cooler water temperatures. As water temperature decreases so does the activity level of a Steelhead. Water temperatures lower than seven degrees Celsius usually demands a sinking line approach. Driven by experience or a simple gut reaction a fly fisher may work a run or a pool with a floating line only to return to the head to work the reaches once again with a sinking line. Specific circumstances or recent angler pressure may have put Steelhead off the bite demanding a closer approach from the angler.

THE DAVIE STREET HOOKER

HOOK: Partridge salmon, No. 2 to 2/0

THREAD: Hot pink

TAG: Flat silver mylar tinsel

UNDERBODY: Silver tinsel or mylar

BODY: Overlapping wraps of hot pink, transparent plastic or mylar

HACKLE: Red hackle fibres over white polar bear hair

UNDERWING: White polar bear hair

WING: White polar bear with overwing of hot pink polar bear and crystal hair

INSTRUCTIONS: Obtain some fluorescent pink kids handlebar streamers from a bicycle shop. Cut the streamers lengthwise to a suitable width. With the thread hanging at the bend and a base of thread on the hook, tie in a length of flat mylar tinsel, wind forward six wraps, then back to create a tag. Wind the thread to the eye, leaving space for the head. Here, attach flat embossed tinsel and wind in close turns, covering the base of the plastic, then back forward, creating a double layer of tinsel. Wrap the plastic over top, slightly overlapping each previous wrap. Align the tips of a small clump of polar bear hair and tie in under the hook, then add a small segment of red hackle over it to complete the throat. The sparse wing is polar bear under a small amount of pink synthetic, tips aligned. Form a head, whip finish and glue.

THE SQUAMISH POACHER

MARCH 1983

During the stark cold of winter coaxing Steelhead to the fly is a challenge. Late run winter fish are single minded, refusing to chase the fly as they work their way upstream to spawn. Water temperatures are stacked against the angler. Steelhead metabolisms are slowed and they hover just above the bottom in the deeper runs and pools within the river. Getting the fly down and deep is paramount for any measure of success. Typical sunk fly presentations involve an upstream cast, mend and following the drift with the rod. As the fly passes by the angler it is at its deepest, but not necessarily at the required depth. Basic mending skills assist in the decent but in order to sink the fly anglers should consider stack mending. Once the upstream cast is made the angler stacks line on top of the fly using a series of wrist rolls reminiscent of a roll cast. Imagine a bucket where the fly has landed and the angler is tasked with placing additional fly line into the bucket. As the line piles onto the water the fly plummets to the bottom unhindered by tension. As the current drifts the fly downstream, the angler continues to add traditional mends as required and either leads or follows the fly with the rod tip.

THE SQUAMISH POACHER

HOOK: #1 to #4 Mustad 7957BX or Eagle Claw 1197G

THREAD: Orange monocord

FEELERS: Mixed hot orange and red bucktail, 50/50

BODY: Fluorescent hot orange chenille, large diameter

LEGS: Hot orange saddle hackle

EYES: Clear green beads, epoxied to 15-pound-test monofilament

CARAPACE: Fluorescent orange surveyor's tape

THE SQUAMISH POACHER

MARCH 1983

INSTRUCTIONS: Glue two clear green beads to pieces of monofilament and cut the surveyors' tape into six-inch by three-eighth-inch strips. Tie in a pair of bead-eyes directly above the point of the hook, hold in place with figure-eight wraps and secure with a drop of head cement. Tie in about 20 bucktail hairs at the bend, pointing slightly downward and extending about an inch. At the same point, attach the saddle hackle by the butt and a length of chenille. Wind the thread to the hook eye. Wrap the chenille to just behind the eyes, then back to the feelers and back to the hook eye, building a triple-layer head before tying off. Wind the hackle in open spirals to the hook eye, tie off and trim the hackle fibres on top. Attach a strip of surveyor's tape at the hook eye then wind the thread carefully back through the hackle to the bend. Pull the tape over the back and tie down at the bend. Wind the thread to just back of the triple layer of chenille, pull excess tape back over and make several thread wraps behind the head. Wind open spirals of thread over both layers of tape to the hook eye and tie off. Trim excess in the shape of a prawn's tail. Make several turns just behind the eye and whip finish.

Rustic life.

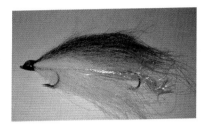

THE SQUAWFISH

MAIN HOOK: 5/0 Mustad stainless
STINGER HOOK: 1/0 Mustad stainless
BODY: Pearl Scent Mylar
BELLY: White and yellow polar bear hair
SIDES: Dark olive and olive polar bear hair
BACK: Dark brown and black polar bear hair
HEAD: Epoxy and sparkles
EYES: stick-on

INSTRUCTIONS: Attach the stinger to the body with a short section of 40-pound test monofilament. Tie the mylar at the tail, wind thread to head followed by mylar body. Build the body by gluing layers of polar bear hair from the belly to the back in the following order: White, pale yellow, olive, dark olive, dark brown and black. Create a large conical head with the tying thread, whip finish and coat with epoxy. When nearly dry stick eyes in place and sprinkle with sparkles. Finish with another layer of epoxy, rotate until dry.

POLAR AURORA BUCKTAIL

MAIN HOOK: Mustad Model 34007, Size 4/0
STINGER HOOK: Mustad Model 34007, Size 2/0
STINGER LINE: A five-inch length of 40-pound-test monofilament
THREAD: Black, 3/0
BODY: Medium holographic mylar
BACK: A small bunch of white polar bear hair, over which a small bunch of pink, dark pink, light blue, dark blue, purple and black polar bear hair in sequence
EYES: Silver eyes

INSTRUCTIONS: Tie the 2/0) hook as your stinger, three inches behind the 4/0 hook. Put a generous amount of head cement on the body of the main hook and add a six-inch piece of holographic mylar, ruffling the tail. Cement each layer of polar bear hair near the eye of the main hook in the above colour order. Use the black thread to create a large head and coat with a generous amount of epoxy, turning often while drying. When nearly dry, put the eyes in place and sprinkle the head with sparkles. Add another coat of epoxy to seal the eyes and sparkles in place. Keep the fly turning until totally dry.

THE GENERAL PRACTITIONER

APRIL 1983

IN RECENT YEARS CLASSIC WET FLY AND SPEY patterns have had to make room for patterns aimed at imitating natural food sources, some of which are popular amongst gear anglers. Ghost or Sand Shrimp are one of the favoured temptations of drift fisherman chasing winter Steelhead. Christened prawn patterns, creations such as the General Practitioner and Raging Prawn have been engineered to duplicate these bugs. Tumbled along the bottom using either a sink tip or shooting head, as water depth and current speed dictate, prawn patterns are ideal choices on runs or rivers that see a fair amount of angler attention owing to their realistic look. Tailouts are good locations to probe even during low water situations of mid winter. The swift surface water masks the slower Steelhead friendly currents present along the bottom while providing a degree of comfort and security. Begin working a tailout by approaching the water's edge with care. Fly fishers first on the water risk trampling resting Steelhead in their attempts to work their way into deeper water. Probe the shallows first and then make progressive casts and swings to sweep the entire tailout.

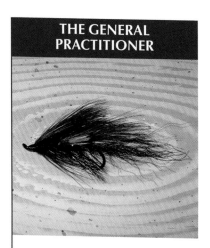

THE GENERAL PRACTITIONER

HOOK: #2 or #4 Mustad 9049

THREAD: Orange or red monocord

FEELERS: Mixed red and orange bucktail

BODY: Pinkish-orange mohair or blended seal fur

RIBBING: Fine gold oval tinsel

EYES: Golden pheasant tippet segments

LEGS: Hot orange saddle hackle

CARAPACE: Layered red-orange golden pheasant neck fibres

THE GENERAL PRACTITIONER

APRIL 1983

INSTRUCTIONS: Tie in 20 to 30 mixed red and orange bucktail hairs in at the bend, followed by a saddle hackle by the butt, gold tinsel, the mohair body material and two small red neck feathers paired, concave sides in, on a horizontal plane. Wind the mohair halfway up the hook and tie, but don't cut. Follow with tinsel then hackle, about three turns each. Now tie in another red hackle feather, horizontally, concave side down to veil the other two. Take two six-fibre segments of tippet and tie in a "V" extending halfway to the head feathers to form the eyes. Head cement everything where it is tied, then continue winding the mohair another quarter of the way up the shank, tie off and follow with the tinsel and hackle, tying them off as well. Cut hackle and tinsel, but leave the mohair. Take a large red feather and tie in as before, veiling the previous feather and the tippet eyes. Wind the mohair to the hook eye, tie off and add another smaller red feather veiling the butt of the previous one. Whip finish and lacquer head to complete.

COQUIHALLA ORANGE

AUGUST 1983

As the fly came under tension a distinct wake formed as it swept across the constant flow of the tailout.

The skies were slightly overcast, water temperature was up and conditions seemed right to draw a summer run to the surface for a fight. The first drift drew no response so after a step downstream the presentation was adjusted to cover a submerged boulder. The fly waked over the boulder as a large shape appeared from below, matching the drift of the skating fly. As the fly reached the shore the Steelhead rolled on the fly but a hookup was not forthcoming. This scenario has jarred many fly fishers leaving them puzzled as to how to coax active Steelhead to the fly. Providing the Steelhead was not stung as it played with the fly it should remain a catchable fish. If after a few drifts the Steelhead does not return or continues to follow but not grab the fly try changing to a wet fly fished damp just below the surface film. Quarter a cast downstream, add mends slowing the fly to the current pace. Apply tension so the fly works through the same arc as the pattern before. Be prepared for a take at anytime through the swing and especially as the pattern hangs below.

COQUIHALLA ORANGE (DARK)

HOOK: Partridge salmon
THREAD: Black
TIP: Fine oval silver tinsel
TAIL: Golden pheasant crest, short orange hackle tips on sides
BUTT: Black ostrich herl
RIBBING: Fine silver wire
BODY: Dark orange floss (rear two-thirds), dubbed hot orange polar bear underfur or seal fur
HACKLE: Hot orange
WING: Hot orange polar bear
OVERWING: Brown mallard
TOPPING: Golden pheasant crest
CHEEKS: Jungle cock

COQUIHALLA ORANGE

AUGUST 1983

COQUIHALLA ORANGE

HOOK: Partridge salmon
THREAD: Black
TAG: Orange floss
TIP: Fine oval gold tinsel
TAIL: Golden pheasant crest
BUTT: Black ostrich herl
RIBBING: Fine silver wire
BODY: Dark orange floss (rear two-thirds), dubbed hot orange polar bear underfur or seal fur
HACKLE: Hot orange
WING: White over hot orange polar bear hair

INSTRUCTIONS: The Coquihalla Orange differs from the dark only in that it has a short tip of fine oval gold tinsel and an orange floss tag; omits the orange hackle tips in the tail, and for the wing, substitutes white polar bear for the brown mallard. Omitted also are the topping and jungle cock cheeks.

THE FREEBIE

SEPTEMBER 1983

Debates amongst seasoned saltwater fly fishers regarding presentation techniques are interesting to listen to. The knowledge and wealth of information these discussions offer to the novice is invaluable. It would appear that each saltwater fly angler creates their own approach, each a success in their own right. Most agree that retrieves should not be brisk. Many prey species meander darting on an occasional basis only. Current and tidal flow has a marked effect on baitfish and their ability to move through the water. So much so that often just holding the baitfish pattern in the flow as though it was treading water can be deadly. In Clayoquot Sound and other areas flush with Needlefish or Sand Lance vertical retrieves work well. Threatened by marauding Coho, Needlefish flee in a vertical fashion, up or down. From a drifting boat make a cast and allow the fly line to sink to a near vertical position. Use a slow to moderate 12-inch strip retrieve and plan for a grab at any time.

THE FREEBIE

HOOK: 2/0 to #2 Mustad 34007
THREAD: White monocord
TAIL: Unravelled silver mylar tubing
BODY: Medium diameter silver mylar tubing
THROAT: Silver mylar strands
WING: Layered polar bear or bucktail, white under magenta, light and dark olive
HEAD: White, lacquered olive green on top
EYES: Painted, yellow with black centre

INSTRUCTIONS: Unravel about an inch of mylar tubing, then slid the tube tail first over the eye of the hook and into place. Attach the thread and bind the mylar down at the hook bend with several wraps and a whip finish. Re-attach the thread at the hook eye, bind down the tubing and clip off the excess. Take a half dozen silver mylar strands long enough to reach the hook point and tie in at the throat. Trim the tail to about half an inch. Prepare the layered wing materials and tie in this order: white, magenta, pale and dark olive. Build a largish conical head and whip finish. Use model lacquer to paint the top of the head olive green. Add large yellow eyes with a black centre.

THE BLACK CREEK

PINK EYE

HOOK: #4 Mustad 34011
THREAD: 6/0 black
RIBBING: Medium silver wire
WING: Bright yellow, chartruese and medium olive polar bear hair
THROAT: Hot orange polar bear hair
INSTRUCTIONS: Lay a base of thread on the shank. Tie in the wire ribbing just up the shank from the bend and wind on about eight evenly spaced turns to the head. The wing is sparse yellow, under chartreuse, under medium olive polar bear hair. It should barely exceed the length of the hook. Tie in a sparse throat of hot orange polar bear hair, about half the length of the wing. Whip finish and coat head with two layers of head cement.

HOOK: #2 to #6 Mustad 34011SS
THREAD: Pink
BODY: Oval tinsel
HEAD (OPTIONAL): Silver, pink or red glass bead head
TAIL/ WING (OPTIONAL): Pink
INSTRUCTIONS: Pinch down the barb with pliers and slide the bead head (optional) onto the hook Tie in a short sparse pink tail (optional) at the bend. Tie in oval tinsel, wind thread to the eye (or behind the bead) and follow by wrapping tinsel forward, tie off. Tie in wing (optional). Whip finish behind the bead and glue or build a head of pink tying thread, whip finish and cover with epoxy.

THE PROFESSOR

DECEMBER/JANUARY 1984

The Halfback and the Carey Special are examples of British Columbia fly patterns that remain popular through generations of fly fishers. For Cutthroat and salmon anglers The Professor is another pattern that has stood the test of time. With the explosion of fly-tying materials both natural and synthetic favourite patterns have benefited from makeovers and facelifts these materials provide. Tied originally with a Red Goose or hackle fibre tail and a yellow floss body The Professor has also benefited from material substitution. Famed B.C. fly fishers Earl Anderson and Jack Vincent replaced the original tail material with a short tuft of red wool and the floss body benefited from the scruffy effects of wool and dubbing. Fly fishers such as Kelly Davison have returned to the red hackle tail but inserted dirty yellow Crystal Chenille for the body. Kelly's version has been christened, "Kelly's Coho Killer" or KCK for short. The Professor and all its permutations are proven Cutthroat and Coho patterns. Perhaps the dirty yellow body suggests a Golden Stone Nymph. Golden Stoneflies inhabit many of the rivers Cutthroat and Coho inhabit.

THE PROFESSOR

HOOK: #2 to #6 Mustad 36890 (for salmon), #6 to #10 Mustad 9671

THREAD: Black

TAIL: Fluorescent red Steelhead yarn

BODY: Medium yellow polar bear underfur or seal fur

RIBBING: Medium gold oval tinsel

HACKLE: Medium brown soft hackle

WING: Rolled mallard flank

INSTRUCTIONS: Dress the hook shank with a layer of tying thread and tie in a short stubby tail of red yarn above the point of the hook, not extending past the bend. Tie in a length of oval tinsel at the tail. Using a dubbing loop painted with sticky wax, wind on a body of polar bear underfur or seal fur, stopping just short of the hook eye. Wind on the ribbing in the opposite direction of the dubbing wraps. Using scissors, trim a bit of the dubbing out at the junction where the wing will be set. Tie in a sparse rolled mallard wing that extends to the tip of the tail and lays fairly close to the shank. Select the softest medium brown hackle you can find, and tie it in. Wind the hackle three or four times, then tie off and trim the fibres from above the lateral line. Form a neat, tapered head of tying thread, whip finish and glue.

THE MICKEY FINN

JANUARY/FEBRUARY 1986

T HE ANNUAL SALMON SPAWNING MIGRATION finds many fly fishers plying local waters driven by visions of battling chrome bright salmon on the fly. Fly fishers are not alone, the annual pursuit of returning salmon clogs local waters with anglers of all types. In populated areas such as the Fraser Valley volume increases at an exponential pace. Competition is fierce. Returning salmon, Coho in particular, witness a myriad of flies and lures. In low water conditions and the slack "frog water" Coho prefer, they become selective and the term, "lock jaw" becomes common banter amongst anglers. Under challenging conditions try small sparse dressings or patterns of a somber nature, sizes six or less. Not the large Christmas tree decorations that may have been successful earlier in the run. The Mickey Finn is an ideal candidate for these conditions and perhaps performs best tied lean and sparse. Don't be shy about varying the Mickey Finn's base colours either. Mix and match the original red and yellow colour scheme to include red and orange or red and chartreuse. Ideal materials for small sparse Mickey Finns include polar bear hair and calf tail.

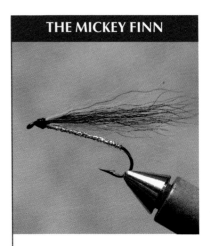

THE MICKEY FINN

HOOK: 2/0 to #6 3X long, nickel-plated

WING: Yellow, red and yellow layered polar bear or bucktail

INSTRUCTIONS: For this pattern simple and sparse is better. Lay a base of thread on a nickel-coated hook. The nickel-coating eliminates the need for a mylar body, making the fly much more durable. Tie in a clump of yellow at the rear of the thread base, then half as much red hair a little further forward, then another clump of yellow equal in size to the first slightly farther forward still. Form a long tapered head out of tying thread, lacquer and set aside to dry.

PETERSON'S LADYBUG

MARCH 1986

Frigid temperatures find Steelhead lethargic and inactive, unwilling to prove their existence. But with each outing new lessons are learned and this is true for Steelhead fly-fishing. During the winter months water levels are low and much of the river is exposed. Pay attention to future obstructions and hollows as these will become future holding areas when the water levels rise in the spring. Observe shallow areas, note all features that in higher water conditions might be prime Steelhead lies. Each year the river changes, channels change course, new holes are scoured, old favourites filled in forever. As an angler's experiences grow areas of the river that hold fish or have the potential to in different circumstances to hold Steelhead. Cached away for future use these walks amongst seemingly barren winter waters pay huge dividends later on in the season as water flow changes and temperatures rise.

PETERSON'S LADYBUG

HOOK: #8 to #14 Mustad 3906B (for trout) 2/0 to #4 Mustad 36890, Eagle Claw 1197B or equivalent (for steelhead)

THREAD: Pale orange or tan

TAIL: Fox Squirrel tail

BODY: Russet brown chenille, sized to suit hook

HACKLE: Pale, muddy, washed-out orange saddle hackle, palmered

BACK: Pale pinkish-orange chenille

RIBBING: Clear monofilament or fine oval gold tinsel

INSTRUCTIONS: Wrap the thread to the bend and tie in a clump of squirrel tail, not stacked, as long as the body for the tail. At the hook bend tie in a length of monofilament or tinsel, a length of pink chenille, the hackle (by the butt, good side toward the eye) and a length of brown chenille in that order. Wrap the thread to the eye and wind on the brown chenille. In wrapping the body, make the turns so that each is just touching the previous one, avoid a tightly wound, compact body. Tie off and follow with the hackle, placing the turns between the chenille wraps. Tie off and trim all the hackle fibres above the lateral line of the fly. Pull the pink chenille over the back and tie off. Wrap the ribbing material over everything, counter-wound to the body materials for strength, pulling the hackle fibres out with a dubbing needle as you go.

Bucktail catch.

ANDERSON'S STONEFLY

OCTOBER 1986

MOST FLY PATTERNS IN USE TODAY WERE CREated with a specific hatch or food item in mind. Suggestive patterns covering a spectrum of presentation or food types are no longer common. Patterns such as the Anderson Stone are tried-and-true designs capable of double shifts on both still and moving waters. Tied originally for Harrison River Cutthroat, this venerable pattern has seduced its fair share of stillwater trout. On the Harrison, an Anderson Stone does a respectable job suggesting the large Golden Stone Nymphs that inhabit the riffles and woody debris. On the numerous interior lakes the Anderson Stone is more suggestive passing itself off as a Dragon Nymph, Scud or something that just looks good to eat. No one knows for sure and trout aren't telling. In the fall on the rich algae type lakes, yellow bodied patterns have become popular due to the angle of the sun and the light levels it creates. The suns rays provide additional attraction to yellow or golden olive-bodied patterns. Golden Yellow Scud or Damsel Nymph patterns are solid autumn performers.

ANDERSON'S STONEFLY

HOOK: #6 to #10 Mustad 9672 or 9671
THREAD: Black monocord
SHELLBACK: Brown wool
ABDOMEN: Pale yellow yarn
THORAX: Same as abdomen
HACKLE: Palmered, dyed medium olive ginger neck hackle

INSTRUCTIONS: Attach a length of brown wool over the barb, on the top of the hook, at the wool's mid-point. With the wool pulled out of the way, attach a length of yellow wool, return the thread to the eye and follow with wraps of the yellow wool to create a uniformly thick body. Pull the brown wool over the back and tie in at the eye. Holding the wool in position, wrap the tying thread back to the bend in tight, widely spaced turns, then make two turns around the bare hook at the end of the abdomen and wrap back to a point two-thirds back of the hook eye. Select a hackle feather where fibres will extend one-and-a-half the hook gape and where fibres over three turns will all be of the same length. Tie in where the thread was left, so that the good side will face forward when wound. Wind thread to the eye using turns as before, follow with three wide turns of hackle and tie off. Whip finish, glue and clip the hackle fibres on the back shortest at the eye and progressively longer to the first wrap.

MY FAVOURITE MINNOW

HOOK: #2 to #8 Mustad 79580
THREAD: Black monocord
TAIL: White calf tail
BODY: 1/8 round pearl mylar tinsel
UNDERBODY: Green, blue or lavender Uni yarn
BELLY: Polar bear hair or white bucktail
OVERWING: Black angel hair
EYES: Plastic stick-on
INSTRUCTIONS: Wind thread to bend and tie in a short sparse tail of calf tail. Cut the tip of the mylar in a bevel so it will ind on flat, tie in at bend. Tie in uni-yarn over mylar. Wind thread to half a centimetre back of hook eye, follow with uni yarn and tie off. Wrap mylar body underbody and tie off. Tie in a sparse belly of polar bear or white bucktail, extending one bend length past the bend of the hook. The wing should be about three to four times as much hair as the belly, and should extend past the bend one hook length. Build a tidy head with thread, whip finish and glue. Coat head with epoxy and allow to dry. Place stick-on eyes and end with another two light coats of epoxy.

SPARKLE LEECH

HOOK: #6 to #12 Tiemco 5263
THREAD: Colour to match
TAIL: Maroon, peacock, brown or green marabou mixed with matching colours of Krystal Flash, Flashabou or Angel Hair
BODY: Twisted maroon and peacock or peacock, brown marron or green Crystal Chenille
HEAD: Gold bead
INSTRUCTIONS: Pinch barb down and lay a even base of thread on the shank. Slide bead onto the hook and secure it with a drop of Krazy Glue. Tie in a tail at the bend of the hook, flash material or Angel Hair first, then marabou on top. The tail should extend at least as long as the body of the fly. Tie in the body materials, twisting together if more than one material is chosen. Wrap to head and tie off. Whip finish behind bead and glue.

MURRAY'S ROLLED MUDDLER

JANUARY/FEBRUARY 1987

THE PURSUIT OF THE NOMADIC COASTAL CUTthroat is an experience unique to many in B.C. Found in coastal lakes, river systems such as the Lower Fraser and along numerous coastal beaches coastal Cutthroat have a loyal following. Coastal Cutthroat are piscivorous, favouring a diet of smaller baitfish and salmon fry. In just about their entire range two prey items coexist, Stickleback and Sculpins. Developed over years of evolution Cutthroat have developed the ability to manage the spiny nature of Sticklebacks making them a preferred food item. In the coastal lakes located on Vancouver Island and the Sunshine Coast a Stickleback imitation is a necessity. Cutthroat often crush the Stickleback within their jaws, casually returning to swallow the cripple. Use a quick line strip to set the hook. Stickleback have also established large populations in coastal rivers such as the Harrison, east of Vancouver. The Harrison and its tributary, the Chehalis, boast a healthy Coho population as well. Immature Coho prey on Stickleback and during the annual Coho spawning migration sparse Stickleback patterns such as the Rolled Muddler lever the latent imprinting of this key food item.

MURRAY'S ROLLED MUDDLER

HOOK: #8 to #12 Mustad 9671 or 9672 Thread Red monocord

TAIL: Rolled mallard flank

BODY: Flat silver mylar tinsel

WING: Rolled mallard flank and sparse deer hair (from collar)

HEAD: Spun, clipped deer hair

INSTRUCTIONS: Form a tail of rolled mallard flank, about three-quarters the hook length, and tie it in at the bend. Tie in the silver wire then the mylar tinsel at the tail. Wrap the tinsel tow-thirds up the hook, then follow with the ribbing wrapped in the opposite direction to reinforce the body. Form a section of rolled wing, nearly as long as the tail, and tie in, ensuring it neither splits nor rolls over. Clip a small bunch of pale deer hair and tie it in by its mid-point, parallel to the hook shank. Allow the hair to flare as the thread is tightened. Repeat with additional small clumps until the eye is reached, then tie off and whip finish. Use a fair amount of thread so that the red thread simulates gills. The hair should be clipped short around the eye, gradually longer to create a oblong bullet or wedge shape, leave a few long hairs at the back of the collar to trail back along the wing. The proper clipping is crucial to the action and appearance of the fly.

THE MAI TAI

AUGUST 1987

Winter run rivers are in a constant state of flux. Water flow, turbidity and depth vary as winter storms approach and pass. In these environments presentation far outweighs pattern. Simple sparse dressings in a variety of sizes work best cutting through the water column in rapid fashion. During high water or spate conditions the flowing water calls many Steelhead in from the sea. Travelling up the shoreline reaches of the river to avoid the surging current fly fishers must approach the fringes with care. Many anglers believing Steelhead to be holding deep stumble and step upon resting fish unaware of their existence. Small feeder creeks offer respite from the currents and Steelhead often hold in the small pockets these tributaries create. One of the most important tactics when scouring a run is to fish the hang down. Popularized by American fly fisher Lani Waller, the hang down is that portion of the presentation where the fly dangles below the angler. Curious by nature, inquisitive Steelhead follow the fly through the swing choosing to slam the fly as hangs in the slack water. If there are no grabs at the end of the drift add a few strips before picking up to recast. A fleeing fly is just the catalyst for a confident take.

THE MAI TAI

HOOK: #2 Eagle Claw 11976 or equivalent

TIP: Flat silver mylar tinsel

THREAD: Orange monocord

TAIL: Silver Doctor blue hackle fibres and short white polar bear hair

BODY: Salmon-orange floss

RIBBING: Fine oval silver tinsel

HACKLE: Mixed white and Silver Doctor blue hackle

WING: Natural white polar bear hair

TOPPING: Three golden pheasant create feathers

CHEEKS: Small jungle cock nails

THE MAI TAI

AUGUST 1987

INSTRUCTIONS: Lay a base of thread on the hook and tie in tip material above the hook point. Wind three wraps back, the three back forward before tying off. Tie in the tail hackle feathers extending to the bend, and overlay with a small, sparse section of polar bear hair, half the length of the existing tail. Tie in the ribbing and then the body floss. Wrap the floss smoothly forward close to the eye, then follow with the ribbing in even wraps. Tie in the mixed hackle next, making no more than three wraps of each, its fibres should fall short of reaching the hook point. Tie in a sparse and fairly short wing supported away from the back by the body floss. Three toppings of golden pheasant crest are followed by small cheeks of jungle cock and a small, neat head. Whip finish and glue.

Tranquil lakes.

THE GREASELINER

SEPTEMBER 1987

THE GREASED LINE APPROACH IS ONE OF THE more well known Steelhead presentations. On the surface this presentation seems straightforward enough, but there is a degree of skill required to work the pattern properly. The greased line method involves making a downstream cast, presenting the fly in a broadside fashion. Done incorrectly a greased line presentation becomes nothing more than a standard wet fly swing. In order to maintain the correct drift and to work the fly with the right degree of tension the fly fisher must be adept at working both the rod and line. Throughout the presentation the rod must be manipulated to either lead or follow the fly. Long, single handed or better yet two handed rods are invaluable. Line management is key, depending upon the current upstream or downstream mends must be made to swim the fly as intended. The fly must work with just enough tension to animate the materials providing a large visible profile for the Steelhead to hone in on. Tall elegant patterns such as Matuka style patterns are ideal for the greased line approach. The greased line method works from bottom to top so long as the angler can get the dynamics of the presentation in order.

THE GREASELINER

HOOK: #2 to #10 light-wire, steelhead/salmon dry-fly

THREAD: Black Flymaster

TAIL: Well-marked mule deer hair

BODY: Dubbed black seal fur or black mohair

HACKLE: Sparse grizzly hackle

WING: well-marked, stiff mule deer hair

INSTRUCTIONS: Tie in the tail above the hook point, extending back about one-quarter the hook length. Using a waxed dubbing loop, apply seal fur and spin the loop to create a rope of dubbing. Wrap forward to just back of the hook eye and tie off, leave sufficient room for the wing and head. Add a few sparse grizzly hackle fibres as a throat. Then add a good-sized, stacked clump of deer hair as a wing, pinching the wing between thumb and forefinger and adding consecutive loose wraps, tightened slowly to ensure that the hair does not roll around the hook shank, then whip finish behind the butts of the hair. Trim the butts to create a small burbler above the eye. Use five-minute epoxy, worked into the hair both in front and behind the wraps, completely impregnating the burbler. Use a pair of toothpicks to flatten and raise the burbler to the desired angle.

NATION'S SILVERTIP

HOOK: #8 3XL

TAIL: Golden pheasant tippets, half body length

BODY: Rear quarter silver tinsel, front portion black dubbing with a good sheen to it Ribbing Embossed silver tinsel, three turns over seal fur

HACKLE: guinea fowl, tied down, two turns

WING: Clipped spun deer hair, bullet shaped with four golden pheasant tippets tied on each side of the wing and allowed to splay out just a bit, tippets should be half of the wing length

GOLDEN PHEASANT

HOOK: #6 3XL

TAIL: Webby, bright red gamecock hackle with a few stiffer fibres included

BODY: Orange-gold chenille or dubbing with a good sheen to it, medium thickness

RIBBING: Medium gold tinsel

HACKLE: Webby, bright red gamecock hackle with a few stiffer fibres included, two turns tied down

WING: Two brown ringneck pheasant saddle hackles, bunched and the length of the body and tail

EYES: Jungle cock (optional)

SILVER THORN

SKYKOMISH SUNRISE

HOOK: #2 Tiemco 9394 or equivalent
THREAD: Light grey
TAIL: Silver Krystal Flash
BODY: Silver tinsel chenille
EYES (OPTIONAL): Bead chain or lead dumbells
UNDERBODY (OPTIONAL): Lead wire

INSTRUCTIONS: Tie eyes in using a series of figure-eight wraps, if you desire them on this pattern, then add a drop of Krazy Glue to secure them. Wrap hook with lead wire, if so desired. Tie in about 12 strands of Krystal Flash at the hook bend, followed by the chenille. Coat the hook shank with head cement and wind on the chenille, crisscrossing twice around the eyes. Form a small head and tie off, finishing with a drop of Krazy Glue.

HOOK: #2 to #12
THREAD: Black
TAIL: Mixed scarlet and yellow hackle wisps
RIBBING: Flat silver tinsel
BODY: Red chenille
COLLAR: Mixed scarlet and yellow saddle hackle
WING: White bucktail

INSTRUCTIONS: Tie in the tail at the bend. Tie on a beveled end of tinsel ribbing, followed by the stripped tip of a section of chenille. Wind thread forward, then chenille, then tinsel in wide, even wraps, tying down each component with a few wraps of thread. Tie in two hackle feathers, scarlet then yellow. Wind two wraps of yellow hackle, tie off. Follow with two wraps of scarlet hackle and tie off. While wrapping hackle, pull fibres back towards point with your fingers. Add a wing of bucktail the length of the tail. Build a neat head, whip finish and glue.

THE PURPLE SPEY

NOVEMBER/DECEMBER 1987

Christened a fish of a thousand casts there are few experiences equalling the initial throb of a large Steelhead at the end of the fly line. Startled by the initial strike and run angler excitement soon switches to anguish as they struggle to subdue a fish measured in grabs and hookups. Battling a large, spirited fish dictates a different approach than a high rod tip until the fish tires. High rod position provides minimal friction benefits the fleeing fish. From a physics perspective a raised rod offers little foot pound resistance to the fish. Once the Steelhead has completed its initial run lower the rod to the side and apply side pressure in the opposite direction to the fish. As the fish alters its course in response to the side pressure counteract its path by flipping the rod to the opposite side. Side pressure allows the angler to exert considerable force, many times that of a traditional high rod position. Side pressure exerts pressure on the steelheads gill plates as the current works for the angler slamming them shut robbing the Steelhead of oxygen. A skilled angler can subdue a large fish in minutes reducing angler fatigue and more importantly, stress on the Steelhead.

THE PURPLE SPEY

HOOK: #2 or #4 Eagle Claw 1197 or equivalent

TAIL: Purple Marabou

BODY: Orange floss

RIBBING: Flat embossed tinsel

HACKLE: Purple saddle hackle

WING: Orange golden pheasant body feather

INSTRUCTIONS: Tie in a clump of purple Marabou at the tail, extending back one hook length. Strip one side of fibres from a dyed purple saddle hackle and tie it in at the tail, followed by a length of ribbing tinsel and orange floss. Wrap the floss up, back and then up the shank again to the eye to produce a smoothly tapered body. Wind the tinsel forward in even wraps to the eye, tie off and trim. Wind the hackle on between the tinsel wraps, spey style. Hold the hackle out of the way and tie on the wing feather, concave side down, so that it extends the length of the body, just slightly over the tail. Use the tying thread to form a smooth head, whip finish and glue.

LORNIE'S WORM

JANUARY/FEBRUARY 1988

Pocket water, regions of a river dotted by large rocks or boulders that create small water pockets are often overlooked by fly fishers. These areas are well oxygenated and offer respite for resident trout, salmon and Steelhead. When fish have seen an onslaught of flies and lures or when conditions are low and warm pocket water is a favourite haunt. Probing and navigating pocket water is a challenge as fly fishers have to struggle to gain a firm foot hold or scale and slide over the many rocks and boulders. A floating line and nine- to 12-foot leader is all that is required in most instances. Pocket water conditions dictate short, accurate casts to scour all potential holding water. The converging currents allow the angler a degree of stealth to approach within close proximity of a target. Flop the fly upstream and raise the rod tip high to permit a deep drift. Make repeated casts before prospecting other areas. Fly anglers must cover the sides, in front and in behind of all obstructions. Indicators can help but in most instances.

LORNIE'S WORM

HOOK: #2 (for Steelhead) or #8 (for trout) 1197 Wright & McGill

UNDERBODY: Heavy lead wire

BODY: Bright or fluorescent pink chenille

HEAD: Hot orange chenille

INSTRUCTIONS: Layer the hook with tying thread to stop the materials from spinning, then wind heavy lead wire from the eye to the bend. Select a medium or large-diameter chenille for Steelhead and a fine-diameter chenille for trout. Start with a piece of pink chenille two-and-a-half times the length of the hook. Tie one end securely to the bend, attach hackle pliers to the other end and twist about a dozen times. Place a dubbing needle at the halfway point on the chenille, and double the hackle pliers back to the hook and tie off. When the dubbing needle is removed the chenille should twist into a rope. If it doesn't or it forms a twisted ball instead, unwrap the end and make adjustments until it twists but hangs straight. Add more pink chenille and cover the lead wire with wraps, leaving a gap near the eye. Using the same method add another twisted tail to the front of the hook using a piece of chenille half the hook's length. Add a few wraps of orange chenille to the gap to form the band. Whip finish to complete.

COHO BLUE

OCTOBER 1988

Fly patterns and presentation techniques for returning salmon and Steelhead are predicated on two basic philosophies. Inducing an aggressive territorial response or awakening a dormant or subdued feeding stimulus. Returning Pacific salmon, such as Coho, are prized quarry for the fly fisher. Early in the run or on those waters that see little or no angler pressure Coho are responsive to the fly demanding little more from the angler other than finding them. Later on in the run or on waters in close proximity to a populated area salmon become fussy, almost uncatchable. Visible Coho adopt an almost nose down posture within a run, actually moving out of the way of both fly line and pattern. Patterns and tippet must be downsized to have any measure of success. Clear fly lines, either slow sinking stillwater or wet tip clear lines, provide an additional advantage when covering spooky Coho. Repeated casts that bring the pattern in from an inoffensive angle from either side or below become the order of battle. Should a Coho break off and follow the fly increase the pace to create further aggravation and aggression. Slowing or stopping the retrieve causes the Coho to lose interest and return to their station.

COHO BLUE

HOOK: #2 1197 DN Wright & McGill
UNDERBODY: .30 lead wire
BODY: Fine diameter mylar tubing
WING: Blue over green polar bear hair

INSTRUCTIONS: Lay a base of thread on the hook and wrap it in lead wire from one end to the other. At the bend of the hook, tie in a length of fine diameter mylar tubing with the string core removed. Wrap it forward over the lead, leaving room for the head. Tie in a clump of green polar bear hair on top of the lead behind the eye of the hook. Wrap it down well and add a drop of glue. Add a slightly larger clump of blue hair on top of the green and tie it securely as well. Whip finish and glue. When the glue is dry, cover the head with silver paint and add some painted eyes.

EGG FLY

NOVEMBER/DECEMBER 1988

EGG PATTERNS ARE SIMPLE TO TIE AND SEEM simple to fish. Bouncing them along the bottom using a sink tip or a floating line and a nine- to 12-foot leader are the two most popular methods. Many anglers using floating lines incorporate small split shot and indicators fishing egg patterns in the same manner as a Nymph, where egg patterns become complicated is size and colour. Steelhead and trout, two popular egg pattern targets, become exacting critics of egg colour. Depending upon the prominent spawning species, tastes can switch many times throughout the season. Pacific salmon produce eggs of varying size and colour. Sockeye produce rich red eggs while light-pink Chum eggs are at the other end of the spectrum. In some waters feeding trout focus upon a specific species eggs at times favouring faded washed out specimens. In addition to colour, size can be critical. All too often anglers opt for the more-is-better theory by choosing large gaudy egg patterns. Try the reciprocal of this approach by downsizing to pint-sized eggs bounced along the bottom rubble.

EGG FLY

HOOK: #6 to 2/0 short shank, down-turned eye
UNDERBODY: Lead wire
BODY: Wool, various colours

INSTRUCTIONS: Note: Most of these patterns have a dominant colour and a touch of a secondary colour (except the Sockeye pattern) as follows: pale pink/orange (Chum), pale orange/red (Chinook/Steelhead), 50/50 orange and red (Sockeye), orange/red (Coho)

Lay a base of thread on the hook and make a few wraps of lead wire on the shank. Smear on some sticky wax and spread short cuttings of wool evenly along the thread in this ratio: two centimetres of the dominant colour, one centimetre of secondary colour, and another centimetre of the dominant colour. Make a dubbing loop and spin it causing the wool to twist into a chenille. Wrap the chenille over the lead, pushing together between wraps. Tie off and glue. Trim to the right size before use.

THE THOMPSON RIVER RAT

HOOK: Tiemco dry fly No.2 to 2/0
THREAD: Red
WING AND TAIL: Lime green polar bear
BODY: White or sulphur yellow caribou or elk
HACKLE: Three or four large hackles to match tail, palmered in separately

CHAN'S SHINER

HOOK: #2 to #8 Tiemco 5263
THREAD: Black, olive or white
BODY: Pearl silver or green pearl mylar tubing
UNDEWING: Pearl Angel Hair
MID-WING: Copper crystal hair
OVERWING: Peacock Angel Hair
EYES: Red or yellow Stick-on

SOL DUC HAIRWING

HOOK: #2 Salmon
THREAD: Red Danville 6/0
BODY: Bright orange floss (rear half), dubbed bright orange seal fur (front half)
RIBBING: Fine oval silver tinsel
WING: Bright orange polar bear hair
THROAT: Black hackle fibres
HACKLE: Yellow, Spey-style

OLIVE STICKLEBACK

HOOK: #6 to #2 Mustad 9672
THREAD: 3/0 black Uni-Thread
TAIL: Olive or white neck hackle
THROAT: White and red marabou
WING: Olive or grey marabou
OVERWING: Rainbow Krystal Flash and opal silver mirage
HACKLE: Black and white grizzly hackle
EYES: Pearl nail polish

Netted catch.

THE STAMP RIVER SHRIMP

JANUARY/FEBRUARY 1989

There has been much discussion over the years regarding fly pattern colours for Steelhead. As this discussion has evolved there seems to be two camps: bright and dark. These two camps seem to based upon the belief that Steelhead take the fly out of aggression or an awakening of a primal feeding instinct. Steelhead, as with other salmonids, have an eye construction similar to humans so their ability to perceive colour is similar. Living in an aqueous environment Steelhead must content with a number of factors including water clarity, depth and available light. Bright patterns featuring red, yellow and orange are popular winter run selections. A brief foray into the theory of light reveals that bright colours might be good selections during the mid-winter day, eliciting a territorial response from returning Steelhead. In the low light hours of morning or evening, however, red, orange and yellow reflect the longest wavelengths of light and as such are the first colours within the spectrum to go, making bright colours poor choices for low light situations. Darker colours, blue and green on the other hand, are the last to disappear. Black, one of the most popular colour choices reflects no light back to the eye making it an ideal choice when a dark silhouette is in order.

THE STAMP RIVER SHRIMP

HOOK: #2 1197 DN Wright & McGill

BODY: Pink and lime green plastic chenille

HACKLE: Dyed green drake mallard flank Feather

WING: Black and white barred wood duck feather

INSTRUCTIONS: Tie one strand of pink and one strand of lime green plastic chenille at the halfway point of the hook. Wrap them forward together to the eye of the hook, forming a banded body. Leave room at the head for the hackle and wing. Strip the barbules from one side of the drake mallard flank feather and tie it in solidly by the butt, then make two or three wraps ahead of the chenille and tie it off. Spread the barbules evenly around the hook. Choose two, well-marker wood duck flank feathers for the wing, moisten with saliva and tie them in, one on either side of the hook. Whip finish and varnish. Finally paint the head with green luminescent paint.

KRILLING YOU SOFTLY

JUNE/JULY/AUGUST 1989

Krill or Euphausids to some are an integral component of a Pacific salmon's diet. Juvenile Coho prey heavily on them at certain times such as mid June when Euphausids gather at the surface to mate. Sea birds can provide clues to potential Krill smorgasbords as they orbit above. Krill feeding birds tend not to be as concentrated as those focused upon bait balls as the Krill tend to spread out more clouding the surface. Favouring dim conditions, mature Euphausids gather at the surface drawing feeding salmon with them. These events provide the best chances of success for the saltwater fly fisher. Light sensitive, mature Euphausids remain deep during the heat of the day. Unlike conventional fast paced baitfish retrieves, slower varied retrieves work best for imitating Krill. When Krill are at the surface use slow sinking clear intermediate lines. Choose a line that places the fly at the same level as the feeding salmon. Use a slow six-inch strip retrieve and try to hop the pattern at the completion of each strip. If the seas are rolling or light conditions might present a barrier to salmon finding the fly, try a series of strips followed by a pause.

KRILLING YOU SOFTLY

HOOK: #4 1197 DN Wright & McGill
BODY: Pink plastic chenille
TAIL: Orange polar bear hair
WHISKERS: Orange polar bear hair
EYES: Bead chain
SHELLBACK: Glow plastic

INSTRUCTIONS: Tie in some orange polar bear hair, about half the hook length, near the head with the tips facing forward. Tie the bead chain eyes in behind the whisker hair butts with several figure-eight wraps, then add a drop of glue to keep them from spinning. Lay a base of thread on the hook as you wind it to the bend. Tie in a few strands of polar bear hair there for a tail, sparser than the whiskers, but about the same length. Tie in the chenille and wind it in even wraps to the eyes, then tie it off and glue. Clip the plastic chenille from the back of the hook shank, then paint the back of the fly and the eyes with luminescent paint.

THE ULTIMATE

JANUARY/FEBRUARY 1990

Seasoned Steelhead fly anglers appear united in a belief, presentation is the critical factor — not fly pattern. A peek into any devoted Steelhead fly fisher's box reveals a select group of favourites tied in a few varied colours and sizes. Small dark flies for low water or pressured fish. Bright patterns for solemn winter runs or large dark patterns for stained, murky waters. Fly pattern behaviour is a key trait. Creative tiers sitting at the vise need to consider how the pattern will swim and behave on the job. Material selection and placement has a marked effect. Long tails or buoyant animal hair such as polar bear or bucktail placed dorsally or along the side ensures a pattern will ride upright as it was intended. Traditional patterns such as the Green Butt Skunk remain popular today due in part to how it tracks during the drift and swing. Weighted patterns are often put into play as well. As with all flies, the location of the weight can be more important than the amount. Many weighted patterns tend to roll upside down altering the performance of the fly. Wise fly fishers can put this trait in their favour by adapting pattern construction. The Boss, for instance looks the same from any angle and owing to its bead chain or dumbbell eyes tumbles and rolls along the bottom with a reduced risk of snags.

THE ULTIMATE

HOOK: #2 or #4 1197 Wright and McGill

BODY: Black chenille or black seal fur

TAIL: Red or orange polar bear hair

BACK: Reddish-brown elk or deer hair

RIBBING: Oval silver tinsel

SIDE WING: Red polar bear hair

TOP WING: Jungle cock eyes or barred wood duck flank feathers

INSTRUCTIONS: Wrap lead wire on a base of tying thread. At the bend of the hook, tie in a piece of oval silver tinsel and a tail of red polar bear hair, as long as the hook shank. Add the body of black chenille or seal fur wrapping to just behind the eye of the hook. Tie a wing of elk or deer hair over the body and part of the tail, then secure it to the body along the back with even wraps of tinsel, tie off. On either side of the hook tie in two sparse, evenly sized clumps of red polar bear hair. Next, add a short wing feather of choice half the length of the hook shank. Whip finish and glue.

SALMON FRY

APRIL 1990

For many baitfish, facing a threat or the fear of one, form dense schools as a primary defensive option. Bait balls are thought of as a saltwater phenomenon but there are bait ball opportunities for British Columbia freshwater anglers to take advantage of as well. Large inland lakes such as Quesnel and Shuswap are Sockeye-driven food factories. Soon after emerging from their gravel redds, Sockeye fry head for the nearest lake where they grow and mature. Rainbows and Dolly Varden char reach huge proportions on a rich diet of Sockeye fry and smolt. Sockeye fry and smolt form huge bait balls that are constantly ravaged by wolf packs of trout and char. Fly fishers on the move should keep their eyes peeled for signs of a feeding frenzy. Surface water appears calm one minute and boiling with slashing Rainbows and fleeing fry or smolt the next. Approach is everything, motor toward the school in a wide arc to avoid spooking feeding fish. An electric motor is ideal for the final approach into casting position. Place the boat upstream from the bait ball, cut the motor allowing the boat to drift beside the fury. Work the edges allowing the fly to fall beneath the bait ball before beginning the retrieve. Use an erratic confused retrieve to suggest an injured or stunned smolt.

SALMON FRY

Note: the materials of this pattern vary depending on what species of fry you are trying to imitate.

HOOK: #4 or #6 2X long (Coho), #2 or #4 3X long (Chinook or Sockeye), #6 or #8 2X long (Chum and Pink) Mustad 9671

TAIL: Orange-brown (Coho), pale-green (Chinook and Sockeye), white or very pale-green (Chum and Pink) Marabou

UNDERBODY: Silver mylar tinsel tube

BODY: Pearlescent Flashabou tube

THROAT: Orange wool (Coho only)

WING: Dyed avocado-green mallard flank feather (Coho), dyed avocado-green teal flank (Chinook), pale-green mallard flank (Sockeye), pale-green mallard flank with a few stands of pale blue hair (Chum and Pink)

SALMON FRY

APRIL 1990

INSTRUCTIONS: Tie in a short clump of Marabou at the hook bend, extending back about half the hook length. Slide on a tube of silver mylar the length of the hook and tie off. Re-fasten the thread at the head and securely wrap the silver mylar. Fold the excess mylar back over the head and tie it down as well, building a bulky head. Tie off the thread, refasten it at the bend and slide on a tube of pearlescent Flashabou. Tie it down at the bend whip finish it there and glue. Re-fasten the thread to the head and tie down the tube. Again fold the excess mylar over the head and tie it down. Add a short, sparse throat of orange wool (for Coho imitations). Cut two strips of flank feather, long enough to cover the back, to form the wing. Wet feather and tie in one piece at a time, concave side down. Build a large head with thread, enough to cover the wool and feather butts. Whip finish and glue, then paint with yellow lacquer when the glue is dry. Add eyes with black lacquer when dry.

Remember the "stringer" days?

THE WOBBLE FLY

JANUARY/FEBRUARY 1991

The importance of size, shape and colour in fly construction is well documented. The motion or behaviour of a pattern is the final critical element. In many instances it is the behaviour of a fly that provides the final trigger for a fish to strike. Material selection is key as many food organisms are translucent, reflective or a combination of the two. Minnow patterns, for instance, must shimmer and shine to suggest natural scales. Body shape is also critical to how a pattern swims and tracks through the water. Baitfish patterns must track true. Creative body sculpture and construction can also make a pattern pitch and hop through the water. The body also provides an acoustic footprint many predator fish hone in on before making the final kill decision. The motion within a fly is another function of behaviour. Supple materials such as rabbit, Marabou and long strand dubbing mixes breathe and flow with the slightest of retrieves. The surreptitious placement of weight further animates mobile materials. Loop knots such as the Non-Slip or Duncan Loop allow for maximum movement allowing the fly to put all aspects of a pattern's behaviour to work for the fly fisher.

THE WOBBLE FLY

HOOK: #4 or #6 1197 DN Wright & McGill or stainless steel O'Shaughnessy or Sproat style

TAIL: Pearlescent Flashabou

BODY: Fat mylar tubing

EYES: Stick-on or painted

THROAT: Orange or red wool

INSTRUCTIONS: Tie in a tail of Flashabou on the curve of the hook so that it hangs slightly downward, extending back about the length of the hook. Paint the shank of the hook with five-minute epoxy and slide on a piece of fat, mylar tubing before it has a chance to set. Tie off either end of the tubing and press the mylar to form a wide, flat body, holding it until the epoxy begins to set and keeps the mylar in place. Shape the mylar so it is flat and slightly follows the curve of the hook around the bend. Add a small piece of orange wool to either side of the throat. Paint or stick eyes on the top and bottom of the body and cover with epoxy.

Cutblocks and cutthroats.

McLEOD'S BURGUNDY WOOLLY BUGGER

HOOK: #2 to #6 Tiemco TMC 5263

TAIL: Burgundy marabou, and pearl crystal hair

BODY: Burgundy ice chenille

HACKLE: #3 grizzly hackle, palmered

RIBBING: Two to three pound monofilament

MUDDLER MINNOW

HOOK: #10 2XL

BODY: Thin fluorescent green floss

WING: Grey webbed sections from a ringneck pheasant tail feather, tied on edge, just a bit longer than the body

HEAD: Clipped spun deer hair, bullet shaped with a sparse ring of hair tips behind the head and trimmed of the top and bottom of the fly

CATHY'S COAT

HOOK: #4 to #6 Mustad 34007 or 34011SS

THREAD: Silver or red

TAIL: Pink Krystal Flash

BODY: Silver mylar and pink Edgebright

WING: Sparsewhite polar bear hair and silver mylar Flashabou

THROAT: Sparse orange saddle hackle

GLENNIE'S PEARLESCENT MICKEY FINN

HOOK: #6 Mustad 3123

THREAD: Red

BODY: Silver tinsel

UNDERWING: Yellow polar bear hair, bucktail or calf tail

WING: Red polar bear hair, bucktail or calf tail

OVERWING: Pearlescent white Flashabou

THE TUBE FLY

MARCH 1991

Although not as common as traditional flies, Tube Flies offer advantages and options worthy of consideration for all aspects of fly-fishing. In B.C., tube flies have become popular with salmon and Steelhead anglers. Stillwater fly fishers on the whole have yet to embrace Tube Flies. Saltwater fly fishers find Tube Flies practical in the caustic environments in which they work their flies. Should a hook rust and begin to perish it is just a matter of hook replacement. The body of the fly remains intact ready to withstand the ravages of many fish. Varying the tube base also provides a degree of versatility. Plastic tubes such as those found in Q-Tips swabs sink at a reduced rate. Conversely metal tubes provide inherent mass allowing fly fishers the opportunity to explore deeper or swifter areas with their patterns. Within the metal tube spectrum various metals and alloy tubes sink at different rates further adding to the options. At the end of the day the individual angler makes their final selections based upon the specific conditions and challenges they face.

THE TUBE FLY

HOOK: No. 2 or 4 treble or short shank 2/0 single

BODY: Q-tip with silver mylar and Flashabou tubing over. Coated with five-minute epoxy

WING: Polar bear hair

EYE: Painted or commercially made eyes

INSTRUCTIONS: Start by covering the plastic tube with silver mylar, either flat or tubular. Tie off an trim, then slide on a section of pearlescent Flashabou tubing. Leave the tail as long as you want the fly to be but trim the head and tie off. Make as many bodies as you wish because the next step is to cover the Flashabou tubing with clear, five-minute epoxy. This protects the fly body from the sharp teeth of the salmon. Don't worry if the epoxy settles to one side of the tubing because this makes a translucent belly and helps to keep the fly upright. When the glue is set, tie in the longest polar bear hair you can find. I start with some white hair, then cover it with blue and finish with some green on top. Using coarse tying thread, tie and glue each clump of hair separately. To finish the fly, I make a head of epoxy, trim it flat with a razor blade, then paint on an eye, or use a commercially made one. To protect the eye, cover it with epoxy.

THE EXCITER

APRIL 1991

ONE OF THE MOST CHALLENGING ASPECTS OF fly fishing for Steelhead is coaxing dour disinterested fish into grabbing the fly. Steelhead that have been in the river for a prolonged period or late winter run with an apparent spawn only mentality complicate an already challenging endeavour. Inactive Steelhead will not move to take the fly and the fly fisher must bounce the fly down their throats to have any measure of success. Presentations must be deliberate and deep in order to succeed. In deep pools, this means heavy sink tips or shooting heads. Keep the leader short to keep the fly in range of the bottom. Areas home to resting Steelhead such as the head of a pool or in a deep run must be scoured in a surgical fashion. Six inches left or right make all the difference. Fly patterns also enter the equation. Loud mobile patterns that elicit an aggressive territorial response are a popular approach. On pressured waters where the constant barrage of flies and gear slams Steelhead mouths, downsizing to less offensive pattern is worthwhile strategy to consider.

THE EXCITER

HOOK: #2 or #4 1197 DN Wright & McGill

THREAD: Heavy black nylon

TAIL: Red Flashabou

BODY: Loosely dubbed black seal fur

EYES: Golden pheasant neck feather with a "V" cut out

WING: Orange or red golden pheasant body

INSTRUCTIONS: Make two turns of dubbed seal fur at the bend of the hook and tie off. Tie in four or five clumps containing five strands each of red Flashabou at various points around the hook shank just in front of the seal fur so it flares out around the hook. Use a dubbing loop to add more black seal fur for four or five wraps, then add the golden pheasant eyes. Continue wrapping the dubbed fur up the shank to the eye and tie off. Add a wing of golden pheasant tied in flat, concave side down. Whip finish and cement the head.

CUTTHROAT ON THE FLY

SEPTEMBER/OCTOBER 1992

Standing ankle deep on a favourite Cutthroat beach, the day looks promising. The surface calm breaks with the subtle roll and odd vicious slash of feeding Cutthroat. Fly line is stripped, a cast is made and the fly plops gently into the ring of the most recent rise. Anticipation of the first take is almost overwhelming. After 10 fruitless minutes it is painfully apparent, Cutthroat are fickle today demanding alternative strategies for success. As with all feeding situations, make an honest effort to determine what might be on the dinner plate. Look amongst the bottom rubble. Are Amphipods or Sculpins darting and scurrying about? Are Stickleback hovering beneath the flotsam that dots the surface? If nothing becomes apparent choose a flashy pattern with varied features for maximum appeal. Key features include prominent eyes, a bit of sparkle or flash and the colour yellow. If minnow patterns aren't working try a non-conformist approach with an Amphipod or Sea Worm pattern, small Leech patterns are great for this. Perhaps a pattern that gurgles along the surface might draw attention. Downsizing patterns is another worthwhile strategy. Mix and match retrieves from slow to fast and don't stick too long with a specific approach or pattern.

CUTTHROAT ON THE FLY

HOOK: No. 10 or 12, 2X long, partridge niflor grey shadow

THREAD: Silver mylar-covered nylon (Wright's metallic thread) (Milniycke — purchased at fabric stores)

BODY: Silver thread underneath Pearlescent Flashabou tubing

UNDERWING: Mixed polar bear hair: blue, green, yellow, red, pink

OVERWING: Grey, bared mallard flank feather

HEAD: Silver mylar thread

CUTTHROAT ON THE FLY

SEPTEMBER/OCTOBER 1992

INSTRUCTIONS: Use silver mylar thread and wrap the complete shank of the hook to the bend. The hook shank must be covered because Flashabou tubing will reflect whatever material it covers. Slide on a tube of Flashabou and fasten it near the bend of the hook with a whip finish. Add a spot of glue here. At this point you can continue the silver thread over the Flashabou tubing, as a form of silver rib, or cut the thread and refasten it near the end of the hook. Take four of five strands each of blue, green, yellow, red and pink polar bear hair, and align the tips in a stacker. Polar bear hair is slippery so fasten it tightly with the tips half a hook-length behind the bend of the hook. Cut a one-centimetre-wide section of grey, barred mallard and wet it with saliva. Fold it over the polar bear wing and fasten it with three or four wraps of thread. Don't worry if the mallard flank breaks into several sections, it will do so anyway as soon as you use it. Trim the butt ends of the hair and feathers and cover the head with the silver thread. Whip finish and glue. The silver mylar thread makes a bulkier head than normal, but this is the intention.

PRINCE GEORGE, B. C.

SYNTHETIC SHINER

HOOK: #2 to #8 Tiemco 800S or 9394
THREAD: White
BODY: None
UNDERWING: Gold Pearl angel hair
SIDES: Holographic Flashabou
MID-WING: Hot Orange Dan Bailey's hi-vis and olive Ultra Hair
OVERWING: Peacock angel hair
TOPPING: Eight to 10 strands of peacock herl
GILLS: Red Crystal hair
EYES: Yellow stick-on

INSTRUCTIONS: Attach thread behind the hook eye and tie in the under wing of Angel hair with a few strands of Holographic Flashabou tied along the sides. Tie the hot orange mid-wing over top, mix in a few strands of copper Flashabou. Next comes the over wing of peacock Angel hair topped with a few strands of peacock herl. Tie in a short beard of red Crystal hair under the hook shank to represent gills. Build up a large conical head with tying thread then whip finish. Colour the top of the head with an olive marker, stick on the eyes and cover with two coats of epoxy.

STEELHEAD BEE

HOOK: Mustad 38960
TAIL: Squirrel tail
BODY: Dubbed brown and yellow seal fur in alternating bands
WINGS: Squirrel tail, cocked forward
HACKLE: Brown hackle

INSTRUCTIONS: Tie in a stubby tail of squirrel hair that extends back the length of the bend past the hook. Then arrange alternating yellow and brown sections of body material on a dubbing loop to mimic the bands of colour on a bee's body, wind to just back of the eye. Tie in a wing that extends the length of the hook, then tie in a sparse hackle feather. Cock wing forward of the eye and make two wraps of hackle behind the wing to hold it in position. Whip finish and glue.

BLACK NINJA

MARCH 1994

Not all beaches make ideal Searun Cutthroat haunts. Fly fishers seeking Cutthroat hideouts should limit their searches to shallow gradient beaches featuring eel grass, small gravel or pebbles and a near by creek. Oyster and Mussel beds are additional features to keep an eye out for. Searun Cutthroat patrol these beaches ambushing Stickleback and Sculpins, aquatic Worms and Amphipods. Many popular Cutthroat beaches, such as those on Vancouver Island, the Oyster River, Black Creek and Bazan Bay are noted for fussy Cutthroat. Cutthroat roaming these areas have witnessed numerous flies and presentations and have become exacting critics, proving difficult to dupe. Common searun Cutthroat practice leans towards use of slender flashy minnow patterns. Fly fishers faced with fickle Cutthroat should swap their minnow patterns for smaller non traditional patterns. Patterns such as the Black Ninja catch Cutthroat off guard, increasing the odds of success.

BLACK NINJA

HOOK: 8, 2X long hook
TAIL: Black Marabou plume, pearlescent Flashabou
HACKLE: Rooster saddle black feather, medium blue hackle
BODY: Plastic, sparkle chenille

INSTRUCTIONS: Wrap a thin coating of thread on the hook shank and put three wraps of lead wire near the eye of the hook, leaving enough room to finish the head. Hold the lead wire in place with a few wraps of the tying thread, then wind the thread to the bend of the hook. For a tail, fasten on a medium clump of black Marabou plume, about the same length as the hook shank. On either side of the Marabou add one strand of pearlescent Flashabou. For the palmer hackle, tie the black feather in by its tip, at the bend of the hook. Be sure the hackle strands are not too long. For a thinner profile, I prefer to strip the barbules from one side of the hackle. Then, wind the chenille over the body, covering the lead wire completely. It may be necessary to smooth the edge of the lead with thread or dubbing. When the chenille is tied off, wind on the black hackle, making about five or six even wraps. Tie off the black hackle and fasten a medium blue hackle at the head of the hook. The blue hackle's barbules should be slightly longer than those of the black hackle. Make only three wraps of the blue hackle then tie it off and whip finish the head. Add a drop of head cement to complete the fly.

MR. EYES

JUNE 1994

Peering into a fly bin of baitfish patterns is reminiscent of being on stage with thousands of eyes staring back. Prominent eyes on baitfish patterns are an absolute must. Most anglers agree, eyes are a key trigger for predator fish. There are numerous types and styles of eyes available to the innovative fly tier. Stick-on eyes seem to be the front runner. Using clear mono thread, build up a large head to suggest the natural head of the baitfish. Take the sheet of stick on eyes and fold the sheet along the individual row of eyes to cup and shape the eyes. Using a bodkin, place the adhesive eyes on the sides of the head and massage them into position using the thumb and forefinger. Once positioned spiral the clear thread over the eyes for additional security and whip finish. Apply a coat of two-ton epoxy to the entire head. Place the fly on a rotary drying wheel to ensure a uniform finish. Try different coloured eyes for different effects, for instance red eyes for an injured baitfish.

MR. EYES

HOOK: #8 to #1 2X or 3X long shank

THREAD: Monocord

UNDERBODY: White floss or wool (for Flashabou body only)

BODY: Tube of silver mylar or Flashabou

(WING): White Phentex

HEAD: Spun white acrylic wool

EYES: Pre-made plastic

INSTRUCTIONS: Wind thread to the bend and slide a tube of silver mylar of Flashabou onto the hook to the bend. If using Flashabou, a smooth underbody of white floss or wool must be made on the shank before the body is added. Tie down mylar at the bend, whip finish, glue and remove thread. Refasten the thread at the hook eye and tie down the tube there, as well. Wind the thread down the shank about 15 millimetres and attach some white Phentex extending half of the hook's length past the bend. Tie the Phentex more thickly on the top of the fly and very sparsely at the sides and bottom, use a needle to separate the strands for a veil-like effect. Bind down the head and trim. Colour the fly with felt pens, pale blue-green backs and iridescent sides to represent Pink, Chum and Sockeye fry (#8 and larger), dark olive backs with orange fins and faint purplish par markings for Cutthroat and Coho fry.

SQUID

JULY/AUGUST 1995

MOST SALTWATER FLY RODDERS THINK OF Herring and Needlefish patterns as fly box staples. Squid imitations are another worthwhile addition. Chinook have a particular weakness for Calamari. Find concentrations of Squid and, as with other saltwater prey gatherings, salmon are not far away. Squid are a key food source during the winter months. Deepwater residents, Squid are not always within fly fishers range from a presentation perspective. However, under certain conditions, such as low light Squid venture to shallower climes well within the reach of salmon and saltwater fly fishers. Squid are chameleons, capable of varied colour schemes depending upon their surroundings and mood. At depth squid glow in the dark and fly tiers are well advised to include fluorescent or glow in the dark materials such as Flashabou in their designs. Fast sinking or shooting head fly lines are best suited to deepwater Squid presentations. In some instances, 800-grain heads are necessary. Most fly fishers struggle casting such rigs and a chuck and duck approach is advised. Using a fast action fly rod make a minimum of hauls to accelerate the head, no more than three casts. With the running line off the reel on the deck launch the final cast upward at an angle. The running line peels off the floor and out to sea. Form an O-ring with the thumb and forefinger to help guide the line and avoid tangles that rob the cast of valuable distance.

SQUID

TUBE: Hollow plastic stem from a Q-tip

THREAD: Strong, white

TAIL/TENTACLES: Six to eight white or pale pink saddle hackles, prepared with either black or luminescent dots painted on them

EYES: Luminescent beads, placed on a strand of 30-pound-test monofilament burned at the ends to hold beads in place

BODY: Flashabou

HEAD: Dubbed white wool cuttings

SQUID

JULY/AUGUST 1995

INSTRUCTIONS: Prepare eyes and hackle. Use black dots to imitate a single squid, and luminescent dots to simulate a group of tiny krill. Slide a tube from a plastic Q-tip onto a long shank hook with the eye and point cut off, then mount in vice. Fasten thread and make a base of thread along the entire tube. Tie in six to eight prepared saddle hackles onto one end, so they surround the tube, and glue the wraps liberally. Tie in the eyes with a series of figure eight wraps. Then slide a tube of Flashabou over and fasten it behind the eyes. Make a dubbing loop and paint it with sticky wax. Spread a thin layer of short white wool cuttings and spin the dubbing into a wide chenille, wrapping it over the length of the tube to the end. The final stage is adding some clear silicone and working it into the wool. Use your fingers to pull the wool back and flatten it into a smooth body. Trim off any errant pieces or sections and set the tube-fly aside to dry.

CLOUSER'S DEEP MINNOW

HOOK: #4 to 2/0 Tiemco 9394 or equivalent straight-eye hook
THREAD: 3/0 to 6/0
EYES: Lead barbells
BELLY: White deer, polar bear, goat or synthetic hair
SIDES: Krystal Flash, colours vary
BACK: Deer, polar bear, goat or synthetic hair, colours vary

INSTRUCTIONS: Start with the hook point down in the vice and the thread one-quarter centimetre back of the hook eye. Make six tight wraps towards the bend, then three back towards the eye to centre it on the wraps. Tie in the lead eyes using figure-eight wraps, half-hitches and a drop of Krazy glue. Lay 20 strands of white belly material along the length of the hook between the eyes and secure with two wraps behind the eyes and a half-hitch in front of the eyes. Turn the hook point up in the vice and tie in 20 strands of Krystal Flash, followed by 20 strands of back material. The dressings can be any length, but the Krystal Flash should extend just beyond the hair. Keep the dressings sparse.

WEIGH WESTER

HOOK: #1 Mustad 34011SS
THREAD: White
TAIL: White polar bear hair
BODY: Mylar piping
EYES: Red

INSTRUCTIONS: Start thread just behind the eye (on-eighth centimetre) and wind to just ahead of the point (one-quarter centimetre). Bind in a tail of polar bear hair 1-1/2 times the length of the hook, then spiral the thread back to the eye. Cut mylar piping to length and fray one end. Slide the opposite end over the eye, then bind down with one tight wrap and a single half hitch. Trim thread. Push mylar rearward over the shank, turning it inside out. Tease the frayed ends to ensure that they are even on both sides of the hook, make three wraps around the bod yjust above the frayed section, then half hitch twice and trim thread. Add glue to the front and rear portion of the body. Stick eyes on the body just above the point. Cover the body with five-minute epoxy and rotate evenly until dry.

STEELHEAD SKATER

NOVEMBER/DECEMBER 1996

STONEFLIES AND CADDIS ARE TWO STAPLES common to most British Columbia Steelhead rivers. Driven by a latent response imitations of two of the largest members of these insects, the Golden Stone and the October Caddis stir childhood memories of returning Steelhead. Large, bushy dry flies drifted and skated above promising Steelhead lairs draw explosive responses. On some waters, such as the legendary Thompson River, returning Steelhead continue to chase waking flies well into the approaching winter, contradicting theories regarding Steelhead activity and water temperature. In recent years the reincarnation of two-hand rods has refined modern Steelhead fly fishing. The control these 12- to 16-foot rods afford permit anglers to cover water previously inaccessible with single handed rods. In the palms of the experienced, fly fishers can cover water repeatedly that would exhaust the single handed fly rodder in short order. The versatility of the spey rod enables fly fishers to fish bushy reaches or other areas untouched due to backcast considerations. Casts such as the 'double-spey' the 'snap-t' or even the standard roll cast are poetry in motion in the hands of a skilled two handed fly caster.

HOOK: Bartlet-style Partridge #2 to 2/0

BODY: Closed cell foam, deer hair, moose mane

STEELHEAD SKATER

NOVEMBER/DECEMBER 1996

INSTRUCTIONS: Slice a piece of foam slightly longer than the hook. The foam should be wider than it is thick. Put a slice down the centre of the foam, halfway through it. Thoroughly cover the shank of the hook with a tying thread then paint it with a fast-drying glue; not head cement. I like Bond 527. Before the glue sets, press the foam over it so the slice covers the hook. After the glue sets, colour the foam with Pantone pen. I like a yellow belly and a brown back. Refasten the tying thread just ahead of the hook point. Keep the wraps on top of each other and don't tighten too much. The thread should just slightly compress the foam. On either side of the foam, add three of four strands of coarse moose mane, they should flare out like legs. Cement the thread junction. Bring the thread over the back of the foam and make another junction. Fasten the moose mane again along the side of the foam. Then, add a wing of stacked deer hair. The hair should extend just beyond the foam and stay on top. Bring the thread forward again over the back of the foam and make another junction. Fasten the butts of the deer hair down on top of the foam and the moose mane along the sides. The moose mane should now be trimmed to form flared legs on either side. The moose mane prevents the fly from rolling onto its side. After the tying thread is whip finished and glued, the butts of deer hair should be trimmed in a flare to push on water and cause more wake. Soak the deer hair head in silicone glue to form a solid waking head.

Master rod builder.

Fish on the line.

THE SCULPIN

SEPTEMBER/OCTOBER 1998

Not all prey fish are slender and streamlined. Bottom dwellers such as Sculpin are flattened and squat, designed for a life darting and crawling on their pectoral fins amongst the bottom debris. Strip retrieves must match the sprint pause swimming motion of the Sculpin to be successful. Placing patterns on or near the bottom is often a necessity for success and a challenge for the fly fisher. On stillwaters and ocean beaches, the countdown method is a great method for surgically locating the fly in the bottom zone. Current swept rivers and streams provide a different dynamic. Sink tip lines and weighted patterns are perfect moving water companions. Weighting patterns in a traditional manner needs to be reconsidered, as patterns created to ride hook up survive longer. Bead chain or dumbbell eyes secured on top of the hook shank cause the pattern to fish point up owing to the keel effect. Dumbbell eyes can be painted or have eyes added as well, providing additional realism.

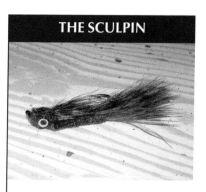

THE SCULPIN

HOOK: #4 or #6 Eagle Claw L52F or equivalent 2X long hook

UNDERBODY: Heavy lead wire and closed cell foam

TAIL: Two soft chicken body feathers, dyed tan, brown or olive

BODY: Dubbed soft body feather plumes or thick untwisted wool, same colour as tail

PECTORAL FINS: Mottled grouse or partridge body feathers

HEAD: Multi-coloured, thick untwisted wool

INSTRUCTIONS: Wrap a layer of tying thread over the hook shank. Make about five wraps of lead wire near the eye and pinch off the excess. From the butt of the wire wrap a foam underbody to the bend and tie off. Wrap the thread slightly around the bend, then tie in a chicken feather on either side of the fly, extending back about half of the hook length, to form the tail. Make a dubbing loop and spin in the body material. Wind the body to the lead wire, but not over it. Select grouse or partridge body feathers to match the colour pattern of the body and tail, tie in on each side, curved side out, flaring away from the body. Dub in a thick head, then trim top and bottom and even up the sides, leaving it flat and broad.

BLACK GENERAL PRACTITIONER

JANUARY/FEBRUARY 1999

Water conditions, in particular clarity, is a prime consideration when choosing an appropriate fly pattern. British Columbia rivers range from brown or glacial spate to the appearance of trying to run underground. To be successful, anglers chasing Steelhead on the fly must be on their toes willing to change and adapt as conditions dictate. When water is clouded and coloured large dark patterns are the primary choice. Swung through promising Steelhead haunts large dark patterns stand out amongst suspended debris and provide an attractive silhouette. Mixture of mobile materials, long strand Marabou or rabbit fur provides additional spice, swaying the odds in the fly fishers' favour. As water lowers and clears fly fishers should consider decreasing their pattern size at a similar rate. During low water conditions large vivid patterns spook wary Steelhead. A 2/0 or larger pattern utilized during freshet or glacial flows needs to be downsized to sizes six or eight to be successful. Fly size can also be altered after a follow or boil on a larger fly. One of the reasons for the refusal could be the dimensions of the original offering. Minimizing the pattern might be just the trick to induce a grab.

BLACK GENERAL PRACTITIONER

HOOK: 5/0 to #4 low-water salmon

TAIL: Black squirrel tail and a small red golden pheasant feather

BODY: Back mohair or wool

RIBBING: Oval silver tinsel

HACKLE: Black cock neck feather

BACK: Two layers of black spade hackles, larger black wood duck or black hen neck feathers, if narrow enough

INSTRUCTIONS: Wrap a three to four-inch piece of lead wire on the hook shank. Tie a clump of black squirrel tail in at the tail include Flashabou or Krystal Flash at your discretion. The tail should extend at least one length of the hook shank beyond the bend. On top of the squirrel, tie in a small red breast feather from a golden pheasant so that it extends about one-quarter the length of the tail. Still at the bend tie in a long piece of mohair from three to six-inches long depending on the hook size.

LOHR'S ALEVIN

MARCH 1999

The alevin is the transitional stage between egg and fry. Translucent and helpless alevin are easy to recognize by their large eyes and prominent egg sack, the alevin's initial food source. Alevin are feeble swimmers, exposed from their gravel haunts they are at the mercy of the elements. An Emerger of sorts, alevin patterns work best early in the season as opportunistic Cutthroat and char are quick to pounce on this easy meal. Floating lines and sink tips are the most popular presentation tools employed by fly fishers. Using an almost Nymph type approach, dead drift the impostor along the bottom suggesting a dislodged alevin. Reach or mend the fly line upstream sinking the fly deep. Areas containing fine gravel are prime alevin locations and should be probed thoroughly. Cutthroat and char scattered amongst the gravel take refuge in redds dug by salmon generations past ambushing unsuspecting alevin and fry. Once the drift is complete allow the pattern to hang in the current below. Trout following the pattern grab it the moment it hesitates. If there are no takers strip the fly upstream to prepare for a recast. Don't be surprised by a jarring interruption prior to picking up the fly rod to cast.

LOHR'S ALEVIN

HOOK: #10 to #14 Tiemco 3761

THREAD: Grey

BODY/TAIL: Light grey Ultra Chenille (Also called Plush or Vernille)

UNDERBODY: Silver holographic tinsel

EGG SAC: Orange egg yarn

EYES: 1.5 millimetre silver prismatic eyes

INSTRUCTIONS: Tie in a piece of Ultra Chenille at the hook bend. Tie in a piece of tinsel. Wrap thread forward and follow by wrapping the tinsel forward to just back of the eye, tie off. Pull the Ultra Chenille over the back of the fly and tie in at the head. Tie in a piece of orange egg yarn under the hook shank, then twist and fold or double it over. If done correctly the yarn should spiral around itself, this technique is called furling. Practice a number of times to get the right number of twists to make the yarn furl properly. Tie off the yarn and build up the head with wraps of thread. Whip finish and place prismatic eyes on either side of the head. Cover the head with clear epoxy. When the epoxy has dried, finish the fly by heating the tail with a flame to taper it.

MAROON BEAD-HEADED WOOLLY BUGGER

BURGUNDY WOOLLY BUGGER

HOOK: #6 to #10 3X long shank
BEAD: Small gold or copper metal bead
TAIL: Maroon Marabou fibres and pearlescent Flashabou
BODY: Maroon plastic chenille
HACKLE: Palmered, short, dyed maroon hackle

INSTRUCTIONS: Pinch down the barb of the hook and slide bead into place (a short layer of thread beneath it will help to hold it in place). Tie in a tail of marabou fibres as long as the hook in at the bend, add two strands of Flashabou to each side of the tail for flash. Attach the chenille body material at the bend, followed by a hackle feather tied in by the tip. Wind the thread and then the chenille forward and tie off behind the bead. Wind hackle forward in five or six evenly spaced wraps. Give an extra two wraps behind the bead, pulling hackle back towards the point after each wrap, tie off. Whip finish behind the bead and glue.

HOOK: Hook: #4 Tiemco 5263
THREAD: 3/0 black Uni-Thread
TAIL: Burgundy marabou and a few strands of Krystal Flash
BODY: Burgundy ice chenille
HACKLE: Palmered grizzly hackle

INSTRUCTIONS: With thread hanging at the bend of the hook, tie in a tail of marabou and Krystal Flash. Tie in hackle by the tip at the same point where the tail was tied. Strip a short section of fluff from the end of a 15 centimetre piece of chenille and tie it in at the bend as well. Wrap thread to just back of the eye, then follow with chenille, tie off. Palmer the hackle in even wraps and tie off at the point where the chenille ends. Whip finish and glue.

CAPITAL PUNISHMENT

SEPTEMBER/OCTOBER 1999

Coho are probably the most pursued of all Pacific salmon species on the fly. Preferring to hold in slow, almost stagnant waters, coho are an ideal quarry for the fly fisher. Christened "frog water" by some, these slow moving Coho climes are ideal candidates for Stillwater or Wet Tip Clear fly lines. Coho that have seen fair numbers of flies shy away from dark coloured sinking lines or sink tips. Prime Coho holding water often features weed beds or sunken debris. Sunken logs and similar rubble are challenging areas for the angler to probe with the fly. Preferred Coho retrieves consist of a six to 12-inch strip retrieve. The overall pace of the retrieve should be moderate, about one strip every second. Start with a slow strip retrieve slowly finishing with a crescendo to induce aggressive Coho to strike. At the end of each strip pop the stripping hand in the same manner as shaking a thermometer to hop and dart the pattern through the retrieve. Visual wakes trailing the fly are common with this approach offering of the most exciting experiences as the Coho tracks down the fly. Takes at the anglers laces are common.

CAPITAL PUNISHMENT

HOOK: #2 to #6 Daiichi 2456
THREAD: Hot pink Danville Plus
TAIL: Hot pink Marabou
BODY: Silver sparkle braid
WING: Electric pink Angel Hair
HACKLE: Hot pink Schlappen

INSTRUCTIONS: Tie in a Marabou feather with lots of web, and even fibres at the tail. Tie in the body material and wrap about three-quarters of the hook shank. Prepare a waxed dubbing loop with Angel Hair cut into about six-centimetre long pieces. Slide the Angel Hair into the loop sideways and twist it tightly. Wrap the loop, ensuring that the tips of the Angel Hair points towards the back of the hook. Tie in the hackle and wrap six times, again ensuring that the fibres point backwards. Build a neat head, whip finish and glue.

THE PURPLE BUNNY

NOVEMBER/DECEMBER 1999

Many fly fishers use one of the varied sink tips available today to search promising steelhead lies. Control is critical, as the fly must be presented at the correct speed. Most fly fishers taking their first forays into steelheading present their flies too fast as result of poor fly line control. Proper fly speed should be at or slightly slower than current speed. Mending is the primary dial fly anglers use to control fly speed. Mends are the upstream or downstream movement of the fly line once the cast is made. In most instances the fly line is reached upstream to both sink and slow the fly. There are times where downstream mends are in order such as working a fly through an inside bend or when the current slows near shore. The trick to mending is being able to reach the line so as not to pull the fly, this is true whether drifting a wet fly for winter Steelhead or a tiny dry fly for sipping trout. A long rod allows for optimum line control and is one of the distinct advantages of a two handed fly rod.

THE PURPLE BUNNY

HOOK: #2 to #4 Wright & McGill 1197 or equivalent 2X long shank hook

TAIL: Dyed purple rabbit fur and purple Flashabou

BODY: Dubbed dyed purple rabbit fur and Flashabou, mixed

WING: Golden pheasant tippet

HACKLE: Dubbed dyed purple rabbit fur

INSTRUCTIONS: Tie in a tail of purple rabbit fur about half the hook length. Top the tail with a few strands of purple Flashabou. Using a dubbing loop and pieces of fur and Flashabou, spin a dubbed body and wind it to within one-half centimetre of the hook eye. Add a wing of orange-golden pheasant tippet with the centre cut out, leaving just a "V" with the distinctive black bars. The wing should lay flat on top. Trim some rabbit fur from the hide about 1.5 centimetres long. Spread the fur evenly along the waxed thread so that the butts are close to the thread and the tips extend to one side. Form a loop and spin, picking out hair as necessary with a needle. Wind this hackle around the neck for three wraps, pulling the hair back after every wrap. Trim, whip finish and glue.

THE BULKLEY BUNNY

SEPTEMBER/OCTOBER 2000

Our Steelhead, particularly winter runs, are amongst the most difficult fish to coax to the fly. As water temperatures drop the challenge only grows deeper. Deep, slow, repetitious presentations are standard procedure. In recent years fly fishers have seen a boon of mobile animated patterns to choose from. Animated materials including Marabou and rabbit are preferred choices. Self motivated flies constructed of these materials pulse and breathe without assistance. In many instances these patterns obtain an aggressive response from seemingly uninterested Steelhead. As a general rule: bright days, bright flies; dark days, dark flies, works for most Steelhead. Don't be afraid to amend this proverb should conditions dictate. A splash of Flashabou or Krystal Flash provides an ideal compliment but avoid the tendency to provide too much of a good thing. Excessive amounts of flash and sparkle can be a Steelhead deterrent. As many Steelhead waters are glacial in nature and subject to freshet, bulky vibrant patterns can be the difference between failure and success. Water changes from the colour of chocolate milk, to glacial blue-green to clear again. Every water condition dictates different adaptations.

THE BULKLEY BUNNY

HOOK: 3/0 to #6 Mustad 36890 Tiemco TMC 7999, Daiichi DA 2421 or DA 2441, Partridge PCS 10/1 or low-water upturned eye

TAIL/BODY: Cerise rabbit strip

WING: Pink Krystal Flash

HACKLE: Purple Schlappen

THREAD: Black 6/0

INSTRUCTIONS: Secure rabbit strip to the shank at the bend leaving a length about two-thirds of the hook shank as a tail. Trim the end into a tapered-"V" for increased action. Wrap the forward portion of the rabbit strip up the shank, stopping after each turn to pick out the fur. Tie off the rabbit strip about seven-eighths up the shank and trim off excess. Tie in six to 12 strands of Krystal Flash on either side of the hook at the neck, not extending past the tail. Select a purple Schlappen hackle with fibres that extend to the hook bend. Tie in and wrap three times before tying off. Whip finish and glue.

MINI GLOW

NOVEMBER/DECEMBER 2000

Mention Cutthroat trout and presentation technique and pattern choice defaults right away to minnow or baitfish. A sound deduction, as nomadic Cutthroat have a natural deposition toward a meal of Stickleback, Sculpin, Chub and salmon fry, to name but a few. Cutthroat cruise coastlines, river and creek mouths, backwaters and sloughs in constant search of hapless fare. Back eddy areas concentrate baitfish, sunken debris provides shelter, and shoreline reaches offer respite from the current and protection of sorts in the shallows. Prolonged exposure to current swept or open areas is not a recipe for continued existence. Fly fishers should explore these areas using minnow patterns as these are prime lies for forage fish and Cutthroat. Retrieves should vary, suggesting startled and passive bait fish. Unaware of a predator in the vicinity baitfish mill about, and hover, their large eyes ever watchful. Try slow six to 12-inch strip retrieves mixed with the odd dash and dart. When plying rivers and streams allow the current to animate the fly as baitfish offer little resistance to the flow. After making a few strips allow the current to pull out the fly line, seesawing the fly cross current. This action provides maximum exposure drawing the attention of any scrupulous Cutthroat.

MINI GLOW

HOOK: Mustad 34007, size 6 to 10
THREAD: Clear monofilament
BODY: Silver or pearlescent Diamond braid
LOWER WING: Polar Ice pearlescent Angel Hair
UNDER WING: Polar White or Silver Grey Shimmer
GILLS: Angler's Choice Shimmer, red
TOPPING: Peacock Angel Head
BEAD: 1/8-inch Silver metal or tungsten Bead
HEAD: Medium pearlescent mylar tubing
EYES: Stick-on eyes; red, yellow or silver

Note: Coat head and eyes with a coating of Devcon Two-Ton epoxy

MINI GLOW

NOVEMBER/DECEMBER 2000

INSTRUCTIONS: Using a pair of pliers, squash the barb of a Tiemco 9394 long-shank streamer hook head on to reduce the risk of accidentally breaking the barb. Once debarbed, slide a 1/8-inch silver metal or tungsten bead onto the hook shank up to the eye. Attach the monofilament tying thread directly behind the bead. Be sure to gain a good foothold with the tying thread, since mono tends to be slippery. Tie in a length of silver or pearlescent Diamond Braid and wind it forward in touching turns to form a neat, slender body covering the rear three-quarters of the shank. Trim the excess Diamond Braid. Prepare a sparse clump of Polar Ice or Pearlescent Angel Hair for the lower wing and secure in place. Stagger-cut the Angel Hair so it extends one half the hook shank, veiling the hook point. Next, tie in a clump of red Shimmer to suggest the gills, a key component to baitfish designs. Trim the gills about one half the distance to the hook point. An underwing of Silver Grey or Polar White tied on the top side of the hook provides subtle movement without bulk. To complete the wings, tie in a topping of peacock Angel Hair. Trim and prune the finished underwing and topping so it is of equal thickness to the lower wing. Keep it sparse since most baitfish appear only as fleeting flashes. Other topping colours to consider include brown, pheasant tail, olive, rainbow and rusty olive. Advance the tying thread in front of the bead to the hook eye. Using a stout pair of scissors, trim a body-length piece of pearlescent mylar tubing and gently remove the thread core. Position the now-hollow tubing over the hook eye so the majority of the mylar is pointing out in front of the hook. Encircle the tubing with two loose, controlled wraps of tying thread. Pull the thread tight around the mylar tubing. Add a few securing wraps for good measure. Move the tying thread behind the bead onto the shoulder of the wing materials. Using a thumb and forefinger, stroke the mylar tubing backwards over the bead turning it inside out in the process. Don't worry if the ends of the tubing fray as this adds pulse and sparkle. Hold the frayed ends in place and bind the tubing onto the wing materials' shoulder directly behind the bead. Whip finish the cement. For the finishing touch, apply a pair of stick-on eyes, red, yellow or silver. Coat the entire head with a smooth coat of Devcon Two-Ton epoxy. The finished head should occupy the front quarter of the fly.

THOMPSON TUBE

JANUARY/FEBRUARY 2003

British Columbian fly fisher Dana Sturn has become an ambassador of both the two-handed spey rod and tube flies for Steelhead. Dana's Thompson Stone or as it is known to most, the Thompson Tube, has proven to be deadly choice for both winter and summer run Steelhead. Dana knots this tube fly onto his tippet more than any other steelhead pattern. A suggestive design, the Thompson Tube imitates the abundant Golden Stone Nymphs that inhabit the majority of British Columbia's hallowed Steelhead streams. Tied in the round, à la the Brooks Stone, the Thompson Tube offers a consistent view no matter the effect of the river. Stirring childhood memories, Steelhead latch on to this fly as it tumbles or sweeps towards them. Dana's favoured presentation method utilizes an across stream cast followed by an upstream mend to sink the fly and setup the drift. Follow the fly with the rod tip until the fly comes under tension and then lead the line with the rod tip swinging the Thompson Tube into the shallows. Takes are typically a soft tap or pluck. Suppress the urge to strike right away, delay the rod lift until the fly line is tight to the fish.

THOMPSON TUBE

TUBE: 1/8 inch diameter air brake hose, 1/2- to 1-1/2-inch in length

HOOK: #2 to #6 Partridge Single Nordic Tube Fly or #4 Tiemco 105

TAIL: Black saddle hackle

BODY: Reddish-brown wire

THORAX (OPTIONAL): Black rabbit dubbing

HACKLE/WING: Black saddle

INSTRUCTIONS: Tube flies are just what they sound like — flies built on tubes. the tubes can be made of plastic, aluminum or brass, depending on the desired sink rates you're after, although we tend to tie most of our flies and then weight them according to our needs just as we would with a traditionally tied fly. You can tie tube flies to look like your regular flies with a traditional dorsal/ventral or top-and-bottom configuration, or you can tie them in the round.

GRANTHAM'S SEDGE

JUNE/JULY/AUGUST 2003

Perhaps the ultimate aphrodisiac for any Steelhead fly fisher is taking a Steelhead on a dry fly, either dead drift or skated. Depending on the fly fisher surface patterns are referred to as skaters, wakers or damp patterns. The common point of reference seems to be whether the fly sits on or in the water. No matter what the definition the goal is the same, lure a Steelhead to the surface for what is hoped is a spectacular strike. The majority of dry fly presentations consist of a downstream quartering cast allowing the fly to drift naturally to start. As the drift continues the fly comes under tension creating an attractive wake. Ideal dry fly runs consist of smooth even flows that allow the fly to sweep through a large arc. Curious Steelhead are often drawn to the fly in the middle of the run following the pattern throughout the swing, choosing to strike as the fly hangs below the angler. Missed strikes occur when a Steelhead smashes the fly during the swing as the pattern is under tension and pulls away from the Steelhead. It takes true angler discipline but upon seeing the rise lower the rod and if possible feed line into the fish to ensure a sound hook up.

GRANTHAM'S SEDGE

HOOK: #4 or #6 Mustad 7957B, 7948A r 94840 down-eye regular length, bronzed, forged or equivalent

HEAD: Stiff 0.065-inch Monofilament

BODY: Black, dark green or golden brown Antron sparkle yarn

WING: Deer hair

INSTRUCTIONS: The thick mono required for this pattern is the type used in electric lawn trimmers. Cut a piece of the mono about half the length of the hook, heat one end over a flame and blunt it, the other end may be beveled with finger nail trimmers for a smoother body. Lay the nylon on the top of the hook shank with the blunt end extending one-quarter of a centimetre past the eye, wrap tightly, tie off and cover with Flex Cement to prevent shifting. From a point just forward of the bend, wrap dubbed body all the way to the end of the blunt extension. Tie in a large clump of stacked deer hair to form a wing, butts extending to blunted extension. The wing tips, about the length of the hook, should flare up and away from the body. Whip finish, Flex Cement the thread and trim body material from around hook eye, if necessary.

CONTRIBUTORS

CONSULTANT
SCOTT BAKER-MCGARVA

Scott Baker-McGarva has been fly casting since he was old enough to wave his arms in the air. Scott was born in Trail, B.C. and grew up fishing storied Interior lakes such as Paul, Jacko, Billy and others under the keen eyes of both his father and grandfather. A direct descendant of famous North Shore pioneer, Austin Spencer (Spent Spinner), Scott spent much of his teen and early 20's steelheading Southern B.C., Alberta's Bow River, and later spent a stint as a bluewater fly-fishing guide in Mexico.

A decade of sport fishing retail, teaching fly tying and various other angling disciplines followed. Scott is now co-owner of the Anglers West Fly and Tackle Shop in Vancouver. Scott has presented at sport fishing shows and guested on televised sport fishing programs.

FOREWORD
BRIAN CHAN

Brian Chan is a fisheries biologist who managed small lake trout fisheries for over 28 years while employed by the province of B.C. Recently, Brian joined the new Freshwater Fisheries Society of B.C., where he is the director of Sport Fishing Development. Brian attended BCIT and obtained a technical diploma in Fish and Wildlife management before completing a Bachelor of Science degree at Simon Fraser University.

Brian has been an ardent fly fisher for the past 35 years with much of his time spent honing skills on small lake trout fisheries. His love of fly fishing led to the development of a secondary career as an author, instructor, lecturer and video producer. Some of Brian's writing credits include *Flyfishing Strategies for Stillwaters* and *Morris and Chan on Fly Fishing Trout Lakes*.

CONTRIBUTORS

CHAPTER INTRODUCTIONS ILLUSTRATIONS CONSULTANT
Ian Forbes

Ian Forbes has been an outdoor writer since the early 1960s and has written for many publications. He has been a regular contributor to BC Outdoors since 1982 and did the Fly of the Month for the magazine from 1987 to 1998.

Ian worked in the B.C. forest industry for over 40 years and is now retired. His work related duties allowed him to visit many places others only dream of. There is not a major river in B.C. that Ian hasn't dropped a fly into. Ian spends over 120 days in the field each year fly fishing as many streams as possible on all parts of the globe. Ian's second love is watercolour painting and illustrating.

CHAPTER FRONTS
Tom Johannesen

Tom Johannesen has been an avid fly-tier and fly-fisherman for over 25 years. He has been published in various fishing magazines as both a writer and photographer and has co-hosted a television program on fishing.

His fishing excursions started on the Lower Mainland and have since taken him throughout British Columbia in search of new and more exciting fisheries to write about.

CONTRIBUTORS

ON THE COVER
JIM MCLENNAN

In the 1970s Jim McLennan began guiding fly fishers on Alberta's Bow River. Since then his writing and photography have appeared in numerous outdoor publications, and he is the author of three books, *Blue Ribbon Bow*, *Trout Streams of Alberta*, and *Fly-Fishing Western Trout Streams*. For 18 years he was co-owner of Country Pleasures, a Calgary fly-fishing store. Jim fly fishes extensively throughout Western Canada.

TECHNIQUE
PHIL ROWLEY

At the age of six Phil Rowley was introduced to coarse fishing in England and has been hooked ever since. For the past 20 years Phil has been fly fishing stillwaters almost exclusively. His love of stillwater fly fishing has taken him all over British Columbia and Washington in the pursuit of trout and char. A former commercial fly tier, Phil has written for almost every major fly fishing publication in North America. His contributions also include books and numerous feature articles on fly fishing stillwaters and stillwater fly patterns. Phil's book *Fly Patterns For Stillwaters* has become a best seller. Phil has traveled Western North America performing at outdoor shows, teaching seminars, speaking to fly clubs and conducting weekend fly-fishing schools. He also has a Website dedicated to fly fishing and fly tying education.

Index

Anderson's Stonefly	140
Arizona Dragon	49
Baggy Shrimp	36
Bead-Body Carey Specials	58
Bead-Head Micro-Mini Leech	52
Beetle	118
Black Creek	135
Black General Practitioner	176
Black Lake Special	27
Black Micro Leech	42
Black Ninja	167
Blood Knot	78
Blue-Winged Olive	53
Bob Giles' River Dragon	29
Bulkley Bunny	181
Burgundy Woolly Bugger	178
Caddis Pupa	22
Callibaetis	33
Callibaetis Mayfly Dun	92
Callibaetis Mayfly Spinner	93
Capital Punishment	179
Case Caddis	19
Cathy's Coat	161
Caverhill Nymph	16
Chan's Shiner	152
Chironomid Larva	26
Chopaka May	97
Cicada	117
Clouser's Deep Minnow	171
Coho Blue	150
Copper Emerger	72
Coquihalla Orange	133
Coquihalla Orange, dark	132
Crayfish	30
Crystal Hair Chironomid	21
Crystal Sedge Pupa	60
Cutthroat on the Fly	164
Damselfly Nymph	17
Dancing Caddis	106
Davie Street Hooker	125
Deer Hair Caddis	100
Deer Hair Stone	105
Deer Wing Backswimmer	51
Dexheimer Sedge	15
Dragonfly	28
Duncan Loop	80
Early Season Bomber	65
Egg Fly	151
Elliott's Iron Blue Humpy	88
Emerald and Copper	59
Ernie's Scud	44
Exciter	163
Fisher's Gnat	119
Flash Chironomid	59
Flashabou Scud	47
Floating Snail	55
Foam Beetle	112
Foam Callibaetis Emerger	101
Freebie	134
Freshwater Snails (floating and sinking)	46
Frostbite Bloodworm	35

Index

General Practitioner 130
Glennie's Pearlescent Mickey Finn 161
Goddard Sedge 112
Gold Bead Marabou Leech (black) 42
Golden Pheasant 146
Golden Stonefly 40
Grantham's Sedge 185
Grasshopper 110
Greaseliner 145
Green Drake 102
Green Drake Emerger 71
Griffith's Gnat 108
Guaranteed 59
Hackle-Wing Mayfly 104
Half and Half 70
Helldiver Leech 20
Improved Clinch Knot 79
Kevin's Orange Shrimp 45
Krilling You Softly 155
Lady McConnell 120
Lohr's Alevin 177
Lohr's Down and Dirty Dragon 54
Lornie's Worm 149
M&M 74
Mai Tai 143
Marabou Damselfly Nymph 56
Marabou Leech 24
March Brown Bead Head 37
Marl Shrimp 18
Maroon Bead-Headed Woolly Bugger 178
McLeod's Burgundy Woolly Bugger 161
McLeod's Chironomid 18
Michael Camp's Caddis 107
Mickey Finn 137
Mikulak Sedge 113
Mini Glow 182
Mo Bradley's Bloodworm 29
Mr. Eyes 168
Muddler Minnow 161
Murray's Rolled Muddler 142
My Favourite Minnow 141
Nail Knot 81
Nation's Silvertip 146
Olive Stickleback 152
Pale Morning Dun 89
Para Pupa 101
Parachute Hackled Fly (Mosquito) 94
Peacock Water Boatman 50

Peterson's Ladybug 138
Pink Eve 135
Plastic Chenille Shrimp 35
Polar Aurora Bucktail 129
Professor 136
Purple Bunny 180
Purple Spey 148
Red Ant 114
Rhyaco 32
Royal Wulff 86
Salmon Candy Sedge, emergent 91
Salmon Candy Sedge, pre-emergent 90
Salmon Fry 157
Sculpin 175
Self Carey 42
Self Carey Special 59
Shaggy Dragon 48
Silver Thorn 147
Simple Sedge 95
Skykomish Sunrise 147
Sol Duc Hairwing 152
Sparkle Leech 141
Squamish Poacher 126
Squawfish 129
Squid 169
Stamp River Shrimp 154
Steelhead Bee 166
Steelhead Skater 172
Stellako Tumbling Stone (adapted) 49
Stimulator 98
Surgeon's Knot 82
Synthetic Shiner 166
Termite 103
Thompson River Rat 152
Thompson Tube 184
Tom Thumb 87
Tube Fly 162
Ultimate 156
Universal Nymph 23
Walhachin Green 89
Water Floatman 63
Weigh Wester 171
Western March Brown 99
Wiggle Fly 39
Wobble Fly 159
Woolly Worm, black 124
Yellow Sally 115